BIOPIRACY

BIOPIRACY

The Plunder of Nature and Knowledge

VANDANA SHIVA

Green Books

in association with The Gaia Foundation

First published in 1998 by
Green Books Ltd
Foxhole, Dartington
Totnes, Devon TQ9 6EB

in association with

The Gaia Foundation
18, Well Walk, Hampstead
London NW3 1LD

First published in the United States by South End Press,
7 Brookline Street, Cambridge, MA 02139

Cover design by Rick Lawrence

Typeset in Bembo at Green Books

Printed web offset by Biddles Ltd, Guildford, UK
on Five Seasons 100% recycled paper

A catalogue record for this book
is available from the British Library

ISBN 1 870098 74 9

CONTENTS

Introduction

Piracy Through Patents:
The Second Coming of Columbus

On April 17, 1492, Queen Isabel and King Ferdinand granted Christopher Columbus the privileges of "discovery and conquest". One year later, on May 4, 1493, Pope Alexander VI, through his 'Bull of Donation', granted all islands and mainlands "discovered and to be discovered, one hundred leagues to the West and South of the Azores towards India", and not already occupied or held by any Christian king or prince as of Christmas of 1492, to the Catholic monarchs Isabel of Castille and Ferdinand of Aragon. As Walter Ullmann stated in *Medieval Papalism*:

> The pope as the vicar of God commanded the world, as if it were a tool in his hands; the pope, supported by the canonists, considered the world as his property to be disposed according to his will.

Charters and patents thus turned acts of piracy into divine will. The peoples and nations that were colonized did not belong to the pope who 'donated' them, yet this canonical jurisprudence made the Christian monarchs of Europe rulers of all nations, "wherever they might be found and whatever creed they might embrace". The principle of 'effective occupation' by Christian princes, the 'vacancy' of the targeted lands, and the 'duty' to incorporate the 'savages' were components of charters and patents.

The Papal Bull, the Columbus charter, and patents granted by European monarchs laid the juridical and moral foundations for the colonization and extermination of non-European peoples. The Native American population declined from 72 million in 1492 to less than 4 million a few centuries later.

Five hundred years after Columbus, a more secular version of the same project of colonization continues through patents and intellectual property rights (IPRs). The Papal Bull has been replaced by the General Agreement on Tariffs and Trade (GATT) treaty. The principle of effective occupation by Christian princes has been replaced by effective occupation by the transnational corporations supported by modern-day rulers. The vacancy of targeted lands has been replaced by the vacancy of targeted life forms and species manipulated by the new biotechnologies. The duty to incorporate savages into Christianity has been replaced by the duty to incorporate local and national economies into the global marketplace, and to incorporate non-Western systems of knowledge into the reductionism of commercialized Western science and technology.

The creation of property through the piracy of other's wealth remains the same as 500 years ago.

The freedom that transnational corporations are claiming through intellectual property rights protection in the GATT agreement on Trade Related Intellectual Property Rights (TRIPs) is the freedom that European colonizers have claimed since 1492. Columbus set a precedent when he treated the licence to conquer non-European peoples as a natural right of European men. The land titles issued by the pope through European kings and queens were the first patents. The colonizer's freedom was built on the slavery and subjugation of the people with original rights to the land. This violent takeover was rendered 'natural' by defining the colonized people as nature, thus denying them their humanity and freedom.

John Locke's treatise on property effectually legitimized this same process of theft and robbery during the enclosure movement in Europe. Locke clearly articulated capitalism's freedom to build as the freedom to steal: property is created by removing resources from nature and mixing them with labour. This 'labour' is not physical, but labour in its 'spiritual' form, as manifested in the control of capital. According to Locke, only those who own capital have the natural right to own natural resources, a right that supersedes the common rights of others with prior claims. Capital is thus defined as a source of freedom that, at the same time, denies freedom to the land, forests, rivers, and biodiversity that capital claims as its own, and to others whose rights are based on their labour. Returning private property to the commons is perceived as depriving the owner of capital of freedom. Therefore peasants and tribespeople who demand the return of their rights and access to resources are regarded as thieves.

These Eurocentric notions of property and piracy are the bases on which the IPR laws of GATT and the World Trade Organization (WTO) have been framed. When Europeans first colonized the non-European world, they felt it was their duty to "discover and conquer", to "subdue, occupy, and possess". It seems that the Western powers are still driven by the colonizing impulse: to discover, conquer, own, and possess everything, every society, every culture. The colonies have now been extended to the interior spaces, the 'genetic codes' of life forms from microbes and plants to animals, including humans.

John Moore, a cancer patient, had his cell lines patented by his own doctor. In 1996, Myriad Pharmaceuticals, a U.S.-based company, patented the breast cancer gene in women in order to get a monopoly on diagnostics and testing. The cell lines of the Hagahai of Papua New Guinea and the Guami of Panama are patented by the U.S. Commerce Secretary.

The natural development and exchange of knowledge has, in effect, been criminalized by the Economic Espionage Act of 1996, which became U.S. law on September 17. The Act empowers U.S. intelligence agencies to investigate the ordinary activities of people worldwide, and considers the intellectual property rights of U.S. corporations as vital to national security.

The assumption of empty lands, *terra nullius*, is now being expanded to 'empty life', seeds and medicinal plants. The takeover of native resources during colonization was justified on the ground that indigenous people did not 'improve' their land. As John Winthrop wrote in 1869:

> Natives in New England, they enclose no land, neither have they any settled habitation, nor any tame cattle to improve the land by soe have nor other but a Natural Right to those countries. Soe as if we leane them sufficient for their use, we may lawfully take the rest.

The same logic is now used to appropriate biodiversity from the original owners and innovators by defining their seeds, medicinal plants, and medical knowledge as nature, as non-science, and treating the tools of genetic engineering as the yardstick of 'improvement'. The definition of Christianity as the only religion, and all other beliefs and cosmologies as primitive, finds its parallel in defining commercialized Western science as the only science, and all other knowledge systems as primitive.

Five hundred years ago, it was enough to be a non-Christian culture to lose all claims and rights. Five hundred years after Columbus, it is enough to be a non-Western culture with a distinctive worldview and diverse knowledge systems to lose all claims and rights. The humanity of others was blanked out then, and their intellect is being blanked out now. Conquered territories were treated as peopleless in the patents of the 15th and 16th centuries. People were naturalized into 'our subjects'.

In continuity with conquest by naturalization, biodiversity is being defined as nature—the cultural and intellectual contributions of non-Western knowledge systems are being systematically erased.

Today's patents have a continuity with those issued to Columbus, Sir John Cabot, Sir Humphery Gilbert, and Sir Walter Raleigh. The conflicts that have been unleashed by the

GATT treaty, by patents on life forms, by the patenting of indigenous knowledge, and by genetic engineering are grounded in processes that can be summarized and symbolized as the second coming of Columbus.

At the heart of Columbus's 'discovery' was the treatment of piracy as a natural right of the colonizer, necessary for the deliverance of the colonized. At the heart of the GATT treaty and its patent laws is the treatment of biopiracy as a natural right of Western corporations, necessary for the 'development' of Third World communities.

Biopiracy is the Columbian 'discovery' 500 years after Columbus. Patents are still the means to protect this piracy of the wealth of non-Western peoples as a right of Western powers.

Through patents and genetic engineering, new colonies are being carved out. The land, the forests, the rivers, the oceans, and the atmosphere have all been colonized, eroded, and polluted. Capital now has to look for new colonies to invade and exploit for its further accumulation. These new colonies are, in my view, the interior spaces of the bodies of women, plants, and animals. Resistance to biopiracy is a resistance to the ultimate colonization of life itself—of the future of evolution as well as the future of non-Western traditions of relating to and knowing nature. It is a struggle to protect the freedom of diverse species to evolve. It is a struggle to protect the freedom of diverse cultures to evolve. It is a struggle to conserve both cultural and biological diversity.

Chapter One

Knowledge, Creativity and Intellectual Property Rights

What is creativity? This is at the heart of the current debates about patents on life. Patents on life enclose the creativity inherent in living systems that reproduce and multiply in self-organized freedom. They enclose the interior spaces of the bodies of women, plants, and animals. They also enclose the free spaces of intellectual creativity by transforming publicly generated knowledge into private property. Intellectual property rights on life forms are supposed to reward and stimulate creativity. Their impact is actually the opposite—to stifle the creativity intrinsic to life forms and the social production of knowledge.

Diverse Creativities

Science is an expression of human creativity, both individual and collective. Since creativity has diverse expressions, I see science as a pluralistic enterprise that refers to different 'ways of knowing'. For me, it is not restricted to modern Western science, but includes the knowledge systems of diverse cultures in different periods of history. Recent work in the history, philosophy, and sociology of science has revealed that scientists do not work in accordance with an abstract scientific method, putting forward theories based on direct and neutral observation. Scientific claims, like all others, are now recognized as arising not out of a verificationist model, but from the commitment of a specialized community of scientists to presupposed metaphors and paradigms, which determine

the meaning of constituent terms and concepts as well as the
status of observation and fact. These new accounts of science,
based on its practice, do not leave us with any criteria to dis-
tinguish the theoretical claims of indigenous non-Western
sciences from those of modern Western science. That it is the
latter that is more widely practiced in non-Western cultures
has more to do with Western cultural and economic hege-
mony than with cultural neutrality. Recognition of diverse
traditions of creativity is an essential component of keeping
diverse knowledge systems alive. This is particularly important in this period of rampant ecological destruction, in
which the smallest source of ecological knowledge and
insights can become a vital link to the future of humanity on
this planet.

Indigenous knowledge systems are by and large ecological,
while the dominant model of scientific knowledge, characterized by reductionism and fragmentation, is not equipped to
take the complexity of interrelationships in nature fully into
account. This inadequacy becomes most significant in the
domain of life sciences, which deal with living organisms.
Creativity in the life sciences has to include three levels:

1. The creativity inherent in living organisms that allows
them to evolve, recreate, and regenerate themselves.

2. The creativity of indigenous communities that have developed knowledge systems to conserve and utilize the rich biological diversity of our planet.

3. The creativity of modern scientists in university or corporate laboratories who find ways to use living organisms to
generate profits.

The recognition of these diverse creativities is essential for the
conservation of biodiversity as well as for the conservation of
intellectual diversity—across cultures and within the university setting.

Intellectual Property Rights and the Destruction of Intellectual Diversity

Intellectual property rights are supposed to reward and provide recognition for intellectual creativity. Yet knowledge and creativity have been so narrowly defined in the context of IPRs that the creativity of nature and of non-Western knowledge systems has been ignored. Theoretically, IPRs are property rights to products of the mind. People everywhere innovate and create: if IPR regimes reflected the diversity of knowledge traditions that account for creativity and innovation in different societies, they would necessarily be pluralistic (also reflecting intellectual modes of property systems and systems of rights), leading to an amazing richness of permutations and combinations.

As currently discussed in global platforms such as GATT and the Biodiversity Convention, or as unilaterally imposed through the Special 301 clause of the U.S. Trade Act, IPRs are a prescription for a monoculture of knowledge. These instruments are being used in order to universalize the U.S. patent regime worldwide, which would inevitably lead to an intellectual and cultural impoverishment by displacing other ways of knowing, other objectives for knowledge creation, and other modes of knowledge sharing.

The TRIPs treaty of the Final Act of GATT is based on a highly restricted concept of innovation. By definition, it is weighted in favour of transnational corporations, and against citizens in general and Third World peasants and forest dwellers in particular.

The first restriction is the shift from common rights to private rights. As the preamble of the TRIPs agreement states, intellectual property rights are recognized only as private rights. This excludes all kinds of knowledge, ideas, and innovations that take place in the "intellectual commons"— in villages among farmers, in forests among tribespeople and even in universities among scientists. TRIPs is therefore a mechanism for the privatization of the intellectual commons and a deintellectualization of civil society. The mind becomes a corporate monopoly.

The second restriction of intellectual property rights is that they are recognized only when knowledge and innovation generate profits, not when they meet social needs. According to Article 27.1, if an innovation is to to be considered an IPR, it has to be capable of industrial application. This immediately excludes all sectors that produce and innovate outside the industrial mode of organization. Profits and capital accumulation are the only ends of creativity; the social good is no longer recognized. Under corporate control, there is a 'deindustrialization' of small-scale production in the informal sectors of society.

By denying the creativity of nature and other cultures, even when that creativity is exploited for commercial gain, 'intellectual property rights' becomes another name for intellectual theft and biopiracy. Simultaneously, people's assertion of their customary, collective rights to knowledge and resources is turned into 'piracy' and 'theft'.

The U.S. International Trade Commission claims that U.S. industry is losing between $100 million and $300 million per year because of 'weak' intellectual property protection in Third World countries.[1] When one takes into account the value of Third World biodiversity and intellectual traditions used freely by commercial interests in the United States, it is the United States—and not countries like India—that is engaged in piracy.

Even though many of the patents in the United States are based on Third World biodiversity and knowledge, it is falsely assumed that without IPR protection, creativity will lie buried. As Robert Sherwood states, "Human creativity is a vast national resource for any country. Like gold in the hills, it will remain buried without encouragement for extraction. Intellectual property protection is the tool which releases that resource."[2]

This interpretation of creativity, as unleashed only when

formal regimes of IPR protection are in place, is a total nega-
tion of creativity in nature as well as the creativity generated
by non-profit motives in both industrial and non-industrial
societies. It is a denial of the role of innovation in traditional
cultures and in the public domain. In fact, the dominant interpretation of IPRs leads to a dramatic distortion in the understanding of creativity, and as a result, in the understanding of
the history of inequality and poverty.

The economic inequality between the affluent industrialized countries and the poor Third World ones is a product of
500 years of colonialism, and the continued maintenance and
creation of mechanisms for draining wealth out of the Third
World. According to the United Nations Development
Program, while $50 billion flows annually from the North to
the South in terms of aid, the South loses $500 billion every
year in interest payments on debts and from the loss of fair
prices for commodities due to unequal terms of trade. Instead
of seeing the structural inequality of the international economic system as lying at the roots of Third World poverty, IPR
advocates explain poverty as arising from a lack of creativity,
which, in turn, is seen as rooted in a lack of IPR protection.

For example, in his book, *Intellectual Property and Economic
Development*, Sherwood relates two stories, one real and one
quite imaginary. In his words, they are meant to draw a contrast between the mindset of ordinary people in countries
with and without effective IPR protection.

> A salesman for a U.S. pump manufacturer, who was a neighbour
> of the author some years ago in upstate New York, noticed
> while visiting customers that a certain type of pressure valve
> would be useful. Although his wife was skeptical, he took time
> at night and weekends to design such a valve and applied for and
> was granted a patent on the design. He placed a second mort-
> gage on his house and later obtained a bank loan, largely on the
> strength of the patent. He created a small business, employed a
> dozen people and contributed to the multiplier effect before the

valve was superseded some 20 years later by other types of valves. The man never thought much about intellectual property. He simply took for granted that he could get a patent and build a business from it.

In Lima, Peru, young Carlos (a fictional proxy for much of the developing world) earns a meagre living welding replacement mufflers under trucks and cars. He thinks of a clamp for simplified muffler installation. His wife is skeptical. Should he spend his nights and weekends to design and develop the clamp? He will need help fabricating a prototype. Should he involve his friend the metalworker? He needs money for metal and tools. Should he use the money saved under the mattress? Should he take a bus across town to ask his sister's husband for a loan? The answer to each question is strongly biased toward the negative by weak intellectual property protection. Without thinking much about intellectual property, his wife, the brother-in-law and Carlos himself each knows from community wisdom that his idea is vulnerable and likely to be taken by others. He cannot take for granted that his idea can be protected.

In this story, lack of confidence that his idea can be protected would in all probability lead Carlos to a negative decision at each of these decision points. If the story of Carlos is multiplied many times across a landscape, that country's opportunity loss is devastating. When an effective protection system becomes a reality, confidence will grow that intellectual assets are valuable and protectable. Then the inventive and creative habit of mind, which is at the heart of an intellectual property protection system, will spread in the minds of people.[3]

Central to the ideology of IPRs is the fallacy that people are creative only if they can make profits and guarantee them through IPR protection. This negates the scientific creativity of those not spurred by the search for profits. It negates the creativity of traditional societies and the modern scientific community, in which the free exchange of ideas is the very condition for creativity, not its antithesis.

Patents as a Block to Free Exchange

There is virtually no evidence that patents actually stimulate invention. Studies, such as Leonard Reich's 1985 *The Making of American Industrial Research*, suggest that patents are used to block other firms from entry into the market. For example, partly as a result of the extension of plant variety protection and the willingness of U.S. courts to extend utility patents to organisms, the number of independent seed companies worldwide has declined markedly in recent decades. Large petrochemical and pharmaceutical giants have extended their corporate reach in the seed market. Such oligopolies often slow down, rather than speed up, the process of invention.

A strong patent system has not been the main reason for economic development, even in industrially developed countries. A study of forty-four large industrial concerns in the U.K., carried out by C. T. Taylor and A. Silberston in 1977, showed that the impact of patents on the rate and direction of invention and innovation was, on the whole, extremely small in all areas examined, with the exception of the secondary (non-basic) chemical industries.

Edwin Mansfield studied U.S. industries on the basis of data from 1981-83. Based on a random sample of 100 firms from twelve industries, patent protection was not essential for electrical equipment, office equipment, motor vehicle, instrument, primary metal, rubber, and textile industries. In another three industries (petroleum, machinery, and fabricated metal products), patent protection was estimated to be essential for the development and introduction of about 10 to 20 percent of their inventions. In pharmaceutical and chemical industries, patents were judged essential for 80 percent of the inventions.

Thus patents are not necessary for developing a climate of invention and creativity. They are more important as instruments of market control. Indeed, the existence of patents undermines the social creativity of the scientific community by stifling free exchange among scientists.

Patents are the strongest form of IPR protection. Wherever patents have been associated with scientific research, the result has been closure of communication. While scientists have never been as open as popular mythology portrays, the threat to scientific communication posed by scientists working with commercial enterprises that seek patent protection is becoming a major cause for concern. As Emanuel Epstein, a noted nucleobiologist, states:

> In the past it was the most natural thing in the world for colleagues to swap ideas on the spur of the moment, to share the latest findings hot off the scintillation counter or the electrophoresis cell, to show each other early drafts of papers, and in other such ways to act as companions in zealous research.
>
> No more. Any UCD [University of California at Davis] scientist with a promising new slant for (crop improvement)... will think twice before talking about it to anyone who is connected with either of the two Davis crop genetic private enterprises—or even with colleagues who in turn might speak to any such person. I know that this type of inhibition is already at work on this campus.[4]

Reflecting on the reduction of openness by scientists in the university-industrial complex, Martin Kenny observes:

> ... the fear of being scooped or of seeing one's work transformed into a commodity can silence those who presumably are colleagues. To see a thing that one produced turned into a product for sale by someone over whom one has no control can leave a person feeling violated. The labor of love is converted into a plain commodity—the work now is an item to be exchanged on the basis of its market price. Money becomes the arbiter of a scientific development's value.[5]

The openness, the free exchange of ideas and information, and the free exchange of materials and techniques have been critical to the creativity and productivity of the research community.

By introducing secrecy to science, IPRs and the associated

commercialization and privatization of knowledge will kill the scientific community, and hence its potential for creativity. IPRs exploit creativity whilst killing its very source. We know that reservoirs that are not replenished soon run dry. Common sense tells us that when roots of a tree are not nourished, it dies. IPRs are an efficient mechanism for harvesting the products of social creativity. They are an inefficient mechanism for nurturing and nourishing the tree of knowledge.

Threats to the Tree of Knowledge

Through subtle processes, the roots of the tree of scientific. knowledge are being starved, even as they are being rapidly exploited and harvested for profits.

The most significant process is what David Ehrenfeld has called 'forgetting'. As certain disciplines and specializations in science spin profits through commercialization, others are neglected, even though they are essential to the foundations of a knowledge system. IPRs lead to the skewing of research to targets of greater commercial interest. As molecular biology becomes a major source of techniques for the biotechnology industry, other disciplines of biology shrivel up and die. We are on the verge of losing our ability to tell one plant or animal from another, and of forgetting how the known species interact among themselves and with their environment.

Earthworms, for instance, are among the species that are crucial to our survival. Agriculture depends on soil fertility, and soil fertility depends heavily on earthworms. They improve the fertility of the soil by depositing their faecal material and increase the permeability of the soil to air and water.

In 1891 Charles Darwin published his last work, the result of a lifetime study of earthworms, in which he wrote:

> It may be doubted whether there are many other animals which have played so important a part in the history of the world, as have these lowly organised creatures.[6]

Yet, as David Ehrenfeld reports, the people who have been trained in earthworm ecology are disappearing:

> At the time of this writing, there is just one actively working scientist who is familiar with the taxonomy of the earthworms of North America. He is at a small private university in Iowa. Another earthworm taxonomist works at a university in Puerto Rico, but she was only recently trained in Spain. A third earthworm taxonomist, trained by his mother, has been working for a post office in Oregon. The fourth, and last, person in North America, north of Mexico, who has expert knowledge of earthworm taxonomy is presently earning a living as a police lawyer in New Brunswick, Canada. There are no more graduate students studying earthworm taxonomy in the United States and Canada. Fifty years ago, at least five American scientists, plus their students, were at work in this field. Nor is the situation different in other parts of the world: Australia, long noted for earthworm research, now has none; the British Museum has ended its earthworm taxonomy, and so on.
>
> The example of earthworms is not atypical. The more advances we make, the more we forget. What use is our expensive technology in a sea of ignorance? [7]

Once priorities shift from social need to potential return on investment, which is the main criterion for commercially guided research, entire streams of knowledge and learning will be forgotten and become extinct. While these diverse fields might not be commercially profitable, they are socially necessary. As a society facing ecological problems, we need epidemiology, ecology and evolutionary and developmental biology. We need experts on particular taxonomic groups, such as microbes, insects, and plants, to respond to the crisis of biodiversity erosion. The moment we ignore the useful and the necessary, and concentrate only on the profitable, we are destroying the social conditions for the creation of intellectual diversity.

Enclosure of the Intellectual Commons

The tree of knowledge also withers from what I have called the 'enclosure of the intellectual commons'. Innovation in the public domain is necessary for the innovation that is privatized by IPRs. The return-on-investment logic that is linked to IPRs, however, fails to replenish the pool of innovations in the public domain. Much of the background research that underlies any patentable development has been publicly funded. Yet the results are often employed in applied research toward patentable discoveries, the rewards of which are appropriated privately.

The movements against TRIPs and patents on life are movements to protect the creativity of nature and of diverse knowledge systems. It is on the conservation of this creativity that our future depends.

Chapter Two

Can Life Be Made?
Can Life Be Owned?
Redefining Biodiversity

In 1971, General Electric and one of its employees, Anand Mohan Chakravarty, applied for a U.S. patent on a genetically engineered *pseudomonas* bacteria. Taking plasmids from three kinds of bacteria, Chakravarty transplanted them into a fourth. As he explained, "I simply shuffled genes, changing bacteria that already existed."

Chakravarty was granted his patent on the grounds that the micro-organism was not a product of nature, but his invention and therefore patentable. As Andrew Kimbrell, a leading U.S. lawyer, recounts: "In coming to its precedent-shattering decision, the court seemed unaware that the inventor himself had characterized his 'creation' of the microbe as simply 'shifting' genes, not creating life." [1]

On such slippery grounds, the first patent on life was granted, and, in spite of the exclusion of plants and animals from patenting under U.S. law, the United States has since rushed to grant patents on all kinds of life forms.

Currently, well over 190 genetically engineered animals, including fish, cows, mice, and pigs, are figuratively standing in line to be patented by a variety of researchers and corporations.

According to Kimbrell:

The Supreme Court's Chakravarty decision has been extended to be continued up the chain of life. The patenting of microbes has led inexorably to the patenting of plants, and then animals. [2]

Biodiversity has been redefined as "biotechnological inventions" to make the patenting of life forms appear less controversial. These patents are valid for twenty years, and hence cover future generations of plants and animals. Yet even when scientists in universities or corporations shuffle genes, they do not 'create' the organism that they then patent.

Referring to the landmark Chakravarty case, in which the court found that he had "produced a new bacterium with markedly different characteristics than any found in nature", Key Dismukes, Study Director for the Committee on Vision of the National Academy of Sciences in the United States, said:

> Let us at least get one thing straight: Anand Chakravarty did not create a new form of life; he merely intervened in the normal processes by which strains of bacteria exchange genetic information, to produce a new strain with an altered metabolic pattern. "His" bacterium lives and reproduces itself under the forces that guide all cellular life. Recent advances in recombinant DNA techniques allow more direct biochemical manipulation of bacterial genes than Chakravarty employed, but these too are only modulations of biological processes. We are incalculably far away from being able to create life de novo, and for that I am profoundly grateful. The argument that the bacterium is Chakravarty's handiwork and not nature's wildly exaggerates human power and displays the same hubris and ignorance of biology that have had such devastating impact on the ecology of our planet.[3]

This display of "hubris and ignorance" becomes even more conspicuous when the reductionist biologists who claim patents on life declare that 95 percent of DNA is 'junk DNA', meaning that its function is not known. When genetic engineers claim to 'engineer' life, they often have to use this 'junk DNA' to get their results.

Take the case of a sheep named Tracy, a 'biotechnological invention' of the scientists of Pharmaceutical Proteins Ltd

(PPL). Tracy is called a 'mammalian cell bioreactor' because; through the introduction of human genes, her mammary glands are engineered to produce a protein, alpha-1-antitrypsin, for the pharmaceutical industry. As Ron James, Director of PPL, states, "The mammary gland is a very good factory. Our sheep are furry little factories walking around in fields and they do a superb job."

While they claim that genetic engineers created the 'biotechnological invention', the scientists at PPL had to use 'junk DNA' to get high yields of alpha-1-antitrypsin. According to James, "We left some of these random bits of DNA in the gene, essentially as God provided it and that produced high yield." In claiming the patent, however, it is the scientist who becomes God, the creator of the patented organism.

Further, future generations of the animal are clearly not 'inventions' of the patent holder; they are the product of the regenerative capacity of the organism. Thus, though the metaphor for patenting is 'engineers' who 'make machines', of the 550 sheep eggs injected with hybrid DNA, 499 survived. When these were transplanted into surrogate mothers, only 112 lambs were born, just five of which had incorporated the human gene into their DNA. Of these, only three produced alpha-1-antitrypsin in their milk, two of whom delivered three grams of protein per litre of milk. But Tracy is the only lamb among the 112 engineered ones to become PPL's 'sheep that lays golden eggs', and produces thirty grams per litre.

One of the characteristics of reductionist biology is to declare organisms and their functions useless on the basis of ignorance of their structure and function. Thus, crops and trees are declared 'weeds'.[4] Forests and cattle breeds are declared 'scrub'. And DNA whose role is not understood is called 'junk DNA'. To write off the major part of the molecule as junk because of our ignorance is to fail to understand biological processes. 'Junk DNA' plays an essential role. The fact that Tracy's protein production increased with the introduction of 'junk DNA' is an illustration of the PPL scientists' ignorance, not their knowledge and creativity.

While genetic engineering is modelled on determinism and predictability, indeterminism and unpredictability are characteristic of the human manipulation of living organisms. In addition to the gap between the projection and practice of the engineering paradigm, there is the gap between owning benefits and rewards and owning hazards and risks.

When property rights to life forms are claimed, it is on the basis of their being new, novel, not occurring in nature. But when it comes time for the 'owners' to take responsibility for the consequences of releasing genetically modified organisms (GMOs), suddenly the life forms are not new. They are natural, and hence safe. The issue of biosafety is treated as unnecessary.[5] Thus, when biological organisms have to be owned, they are treated as unnatural; when the ecological impact of releasing GMOs is called to account by environmentalists, these same organisms are now natural. These shifting constructions of 'natural' show that science, which claims the highest levels of objectivity, is actually very subjective and opportunistic in its approach to nature.

The inconsistency in the construction of 'natural' is well illustrated in the case of the manufacture of genetically engineered human proteins for infant formula. Gen Pharm, a biotechnology company, is the owner of the world's first transgenic dairy bull, called Herman. Herman was bio-engineered by company scientists when still an embryo to carry a human gene for producing milk with a human protein. The milk was to be used for making infant formula.

The engineered gene, and the organism of which it is a part, are treated as unnatural when it comes to ownership of Herman and his offspring. Yet when it comes to the safety of the infant formula containing this bioengineered ingredient extracted from the udders of Herman's offspring, the same company says: "We're making these proteins exactly the way they're made in nature." Gen Pharm's Chief Executive Officer, Jonathan MacQuitty, would have us believe that infant formula made from human protein bioengineered in the milk of transgenic dairy cattle is human milk. "Human milk is the gold standard, and formula companies have added more and more [human elements] over the past 20 years." From this perspective, cows, women, and children are merely instruments for commodity production and profit maximization.[6]

As though the inconsistency between the construction of the natural and novel in the spheres of patent protection and health and environmental protection was not enough, Gen Pharm, the 'owner' of Herman, has totally changed the objective for making a transgenic bull. They now have ethical clearance on the grounds that, by using him for breeding, the modified version of the human gene for lactoferin might be of benefit to patients with cancer or AIDS.

Patenting living organisms encourages two forms of violence. First, life forms are treated as if they are mere machines, thus denying their self-organizing capacity. Second, by allowing the patenting of future generations of plants and animals, the self-reproducing capacity of living organisms is denied.

Living organisms, unlike machines, organize themselves. Because of this capacity, they cannot be treated as simply 'biotechnological inventions', 'gene constructs', or 'products of the mind' that need to be protected as 'intellectual property'.

The engineering paradigm of biotechnology is based on the assumption that life can be made. Patents on life are based on the assumption that life can be owned because it has been constructed.

Genetic engineering and patents on life are the ultimate expression of the commercialization of science and the commodification of nature that began the scientific and industrial revolutions. As Carolyn Merchant has analyzed in *The Death of Nature*, the rise of reductionist science allowed nature to be declared dead, inert, and valueless. Hence it allowed for the exploitation and domination of nature, in total disregard of the social and ecological consequences.[7]

The rise of reductionist science was linked with the

commercialization of science, and resulted in the domination of women and non-Western peoples. Their diverse knowledge systems were not treated as legitimate ways of knowing. With commercialization as the objective, reductionism became the criterion of scientific validity. Non-reductionist and ecological ways of knowing, and non-reductionist and ecological systems of knowledge, were pushed out and marginalized.

The genetic engineering paradigm is now pushing out the last remains of ecological paradigms by redefining living organisms and biodiversity as 'man-made' phenomena.

The rise of the reductionist paradigm of biology to serve the commercial interests of the genetic engineering industry was itself engineered. This was done through funding as well as rewards and recognition.

Genetic Engineering and the Rise of the Reductionist Paradigm of Biology

Reductionism in biology is multifaceted. At the species level, this reductionism puts value on only one species—humans—and generates an instrumental value for all others. It therefore displaces and pushes to extinction all species whose instrumental value to humans is small or non-existent. Monocultures of species and biodiversity erosion are the inevitable consequences of reductionist thought in biology, especially when applied to forestry, agriculture, and fisheries. We call this first-order reductionism.

Reductionist biology is increasingly characterized by a second-order reductionism—genetic reductionism—the reduction of all behaviour of biological organisms, including humans, to genes. Second-order reductionism amplifies the ecological risks of first-order reductionism, while introducing new issues, like the patenting of life forms.

Reductionist biology is also an expression of cultural reductionism, since it devalues many forms of knowledge and ethical systems. This includes all non-Western systems of agriculture and medicine, as well as all disciplines in Western biology that do not lend themselves to genetic and molecular reductionism, but are necessary for dealing sustainably with the living world.

Reductionism was promoted strongly by August Weismann, who nearly a century ago postulated the complete separation of the reproductive cells—the germ line—from the functional body, or soma. According to Weismann, reproductive cells are already set apart in the early embryo and continue their segregated existence into maturity, when they contribute to the formation of the next generation. This supported the idea that acquired traits with no direct feedback from the environment were non-inheritable. The mostly non-existent 'Weismann barrier' is still the paradigm used to discuss biodiversity conservation as 'germ plasm' conservation. The germ plasm, Weismann had earlier contended, was divorced from the outside world. Evolutionary changes toward greater fitness—meaning greater capacity to reproduce—were the result of fortuitous mistakes that happened to prosper in the competition of life.[8]

Weismann's classic experiment a century ago was taken as proof of the non-inheritability of acquired characteristics. He cut the tails off twenty-two generations of mice and found that the next generation was still born with normal tails. The sacrifice of hundreds of mouse tails only proved that this type of mutilation was not inherited.[9]

The proposition that information only goes from genes to the body was reinforced by molecular biology and the discovery in the 1950s of the role of nucleic acid, placing Mendelian genetics on a solid material basis. Molecular biology showed a means of transferral of information from genes to proteins, but gave no indication—until recently—of any transfer in the opposite direction. The inference that there could be none became what Francis Crick called the central dogma of molecular biology: "Once 'information' has passed into proteins, it cannot get out again." [10]

Isolating the gene as a 'master molecule' is part of biological determinism. The 'central dogma' that genes as DNA make proteins is another aspect of this determinism. This dogma is preserved even though it is known that genes 'make' nothing. As Richard Lewontin states in *The Doctrine of DNA*:

> DNA is a dead molecule, among the most non-reactive, chemically inert molecules in the world. It has no power to reproduce itself. Rather, it is produced out of elementary materials by a complex cellular machinery of proteins. While it is often said that DNA produces proteins, in fact proteins (enzymes) produce DNA.
>
> When we refer to genes as self-replicating, we endow them with a mysterious autonomous power that seems to place them above the more ordinary materials of the body. Yet if anything in the world can be said to be self-replicating, it is not the gene, but the entire organism as a complex system.[11]

Genetic engineering is taking us into a second-order reductionism not only because organisms are perceived in isolation of their environment, but because genes are perceived in isolation of the organism as a whole.

The doctrine of molecular biology is modelled on classical mechanics. The central dogma is the ultimate in reductionist thought.

At the very same time that Max Planck, Niels Bohr, Albert Einstein, Erwin Schrödinger and their brilliant colleagues were revising the Newtonian view of the physical universe, biology was becoming more reductionist.[12]

Reductionism in biology was not an accident, but a carefully planned paradigm. As Lily E. Kay records in *The Molecular Vision of Life*, the Rockefeller Foundation served as a principal patron of molecular biology from the 1930s to the 1950s. The term 'molecular biology' was coined in 1938 by Warren Weaver, the Director of the Rockefeller Foundation's Natural Science Division. The term was intended to capture the essence of the foundation's programme—its emphasis on the ultimate minuteness of biological entities.

The cognitive and structural reconfigurations of biology into a reductionist paradigm were greatly facilitated through the economically powerful Rockefeller Foundation. During the years 1932-1959, the foundation poured about $25 million into molecular biology programmes in the United States, more than a quarter of the foundation's total spending for the biological sciences outside medicine (including, from the early 1940s on, enormous sums for agriculture).[13]

The force of the foundation's funding set the trends in molecular biology. During the dozen years following 1953 (the elucidation of the structure of DNA), Nobel Prizes were awarded to scholars for research into the molecular biology of the gene, and all but one had been either fully or partially sponsored by the Rockefeller Foundation under Weaver's guidance.[14]

The motivation behind the enormous investment in the new agenda was to develop the human sciences as a comprehensive explanatory and applied framework of social control grounded in the natural, medical, and social sciences. Conceived during the late 1920s, the new agenda was articulated in terms of the contemporary technocratic discourse of human engineering, aiming toward restructuring human relations in congruence with the social framework of industrial capitalism. Within that agenda, the new biology (originally named 'psychobiology') was erected on the bedrock of the physical sciences in order to rigorously explain and eventually control the fundamental mechanisms governing human behaviour, placing a particularly strong emphasis on heredity. Hierarchy and inequality were thus 'naturalised'. As Lewontin states in *The Doctrine of DNA*:

> The naturalistic explanation is to say that not only do we differ in our innate capacities but that these innate capacities are themselves transmitted from generation to generation biologically. That is to say, they are in our genes. The original social and economic notion of inheritance has been turned into biological inheritance.[15]

The conjunction of cognitive and social goals in reductionist biology had a strong historical connection to eugenics. As of 1930, the Rockefeller Foundation had supported a number of eugenically directed projects. By the time the 'new science of man' was inaugurated, however, the goal of social control through selective breeding was no longer socially legitimate.

Precisely because the old eugenics had lost its scientific validity, a space was created for a new programme that promised to place the study of human heredity and behaviour on vigorous grounds. A concerted physicochemical attack on the gene was initiated at the moment in history when it became unacceptable to advocate social control based on crude eugenic principles and outmoded racial theories. The molecular biology programme, through the study of simple biological systems and the analyses of protein structure, promised a surer, albeit much slower, way toward social planning based on sounder principles of eugenic selection.[16]

Reductionism was chosen as the preferred paradigm for economic and political control of the diversity in nature and society.

Genetic determinism and genetic reductionism go hand in hand. But to say that genes are primary is more ideology than science. Genes are not independent entities, but dependent parts of an entirety that gives them effect. All parts of the cell interact, and the combinations of genes are at least as important as their individual effects in the making of an organism.

More broadly, an organism cannot be treated simply as the product of a number of proteins, each produced by the corresponding gene. Genes have multiple effects, and most traits depend on multiple genes.

Yet the linear and reductionist causality of genetic determinism is held on to, even though the very processes that make genetic engineering possible run counter to the concepts of 'master molecules' and the 'central dogma'. As Roger Lewin has stressed:

Restriction sites, promoters, operators, operons, and enhancers play their part. Not only does DNA make RNA, but RNA, aided by an enzyme suitably called reverse transcriptase, makes DNA.[17]

The weakness of the explanatory and theoretical power of reductionism is made up for by its ideological power as well as its economic and political backing.

Some biologists have gone far in exalting the gene over the organism and demoting the organism itself to a mere machine. The sole purpose of this machine is its own survival and reproduction, or perhaps more accurately put, the survival and reproduction of the DNA that is said both to programme and to 'dictate' its operation. In Richard Dawkins' terms, an organism is a 'survival machine'—a 'lumbering robot' constructed to house its genes, those 'engines of self-preservation' that have as their primary property inherent 'selfishness'. They are sealed off from the outside world, communicating with it by tortuously indirect routes, manipulating it by remote control. They are in you and in me; they created us, body and mind. And their preservation is the ultimate rationale for our existence.[18]

This reductionism has epistemological, ethical, ecological, and socioeconomic implications.

Epistemologically, it leads to a machine view of the world and its rich diversity of life forms. It makes us forget that living organisms organize themselves. It robs us of our capacity for the reverence for life—and without that capacity, protection of the diverse species on this planet is impossible.

Engineering vs. Growing

The capacity to self-organize is the distinctive feature of living systems. Self-organizing systems are autonomous and self-referential. This does not mean that they are isolated and non-interactive. Self-organized systems interact with their environment, but maintain their autonomy. The environment merely triggers the structural changes; it does not specify or direct them. The living system specifies its own structural changes and which patterns in the environment will trigger them. A self-organizing system knows what it has to import and export in order to maintain and renew itself.

Living systems are also complex. The complexity of their structure allows for self-ordering and self-organization. It also allows for the emergence of new properties. One of the distinguishing properties of living systems is their ability to undergo continual structural changes while preserving their form and pattern of organization.

Living systems are also diverse. Their diversity and uniqueness is maintained through spontaneous self-organization. The components of a living system are continually renewed and recycled through structural interaction with the environment, yet the system maintains its pattern, its organization, and its distinctive form.

Self-healing and repair is another characteristic of living systems that derives from complexity and self-organization.

The freedom for diverse species and ecosystems to self-organize is the basis of ecology. Ecological stability derives from the ability of species and ecosystems to adapt, evolve, and respond. In fact, the more degrees of freedom available to a system, the more a system can express its self-organization. External control reduces the degrees of freedom a system has, thereby reducing its capacity to organize and renew itself.

Ecological vulnerability comes from the fact that species and ecosystems have been engineered and controlled to such an extent that they lose the capacity to adapt and evolve.

Chilean scientists Humberto R Maturana and Francisco J. Varela have distinguished two kinds of systems—autopoietic and allopoietic. A system is autopoietic when its function is primarily geared toward self-renewal. An autopoietic system refers to itself. In contrast, an allopoietic system, such as a machine, refers to a function given from outside, such as the production of a specific output.[19]

Self-organizing systems grow from within, shaping themselves outwards. Externally organized mechanical systems do not grow; they are made, put together from the outside.

Self-organizing systems are distinct and multidimensional. They therefore display structural and functional diversity. Mechanical systems are uniform and unidimensional. They display structural uniformity and functional one-dimensionality.

Self-organizing systems can heal themselves and adapt to changing environmental conditions. Mechanically organized systems do not heal or adapt; they break down.

The more complex a dynamic structure is, the more endogenously it is driven. Change depends not only on its external compulsions, but on its internal conditions. Self-organization is the essence of health and ecological stability for living systems.

When an organism or a system is mechanically manipulated to improve a one-dimensional function, including the increase in one-dimensional productivity, either the organism's immunity decreases, and it becomes vulnerable to disease and attack by other organisms, or the organism becomes dominant in an ecosystem and displaces other species, pushing them into extinction. Ecological problems arise from applying the engineering paradigm to life. This paradigm is being deepened through genetic engineering, which will have major ecological and ethical implications.

Ethical Implications of Genetic Engineering

When organisms are treated as if they are machines, an ethical shift takes place—life is seen as having instrumental rather than intrinsic value. The manipulation of animals for industrial ends has already had major ethical, ecological, and health implications. The reductionist, machine view of animals removes all barriers of ethical concern for how animals are treated to maximize production. Within the industrial livestock production sector, the mechanistic view predominates.

For example, a manager of the meat industry states that:

> The breeding sow should be thought of as, and treated as, a valuable piece of machinery, whose function is to pump out baby pigs like a sausage machine.[20]

Treating pigs as machines, however, has a major impact on their behaviour and health. In animal factories, pigs have to have their tails, teeth, and testicles cut off because they fight with each other and resort to what the industry calls 'cannibalism'. Eighteen percent of the piglets in factory farms are choked to death by their mother. Two to five percent are born with congenital defects, such as splayed legs, no anus, or inverted mammary glands. They are prone to disease, such as 'banana disease' (so named because stricken pigs arch their backs into a banana shape) or Porcine Stress Syndrome.

These stresses and diseases are bound to increase with genetic engineering. Already, the pig with the human growth hormone has a body weight that is more than its legs can carry.

The issues of health and animal welfare are intrinsically related to the ecological impact of the new technologies on the capacity of self-regulation and healing. The issue of intrinsic worth is intimately related to the issue of self-organization, which is also, in turn, related to healing.

In the making of the organism, the multiplying cells seem to be instructed as to their respective destinies, and they become permanently differentiated to compose organs. But the instructions or pattern for making the whole structure remain somehow latent. When a part is injured, some cells become undifferentiated in order to make new, specialized tissues.[21]

Thus, there is a self-directed capacity for restoration. The faculty of repair is, in turn, related to resilience. When organisms are treated as machines, and manipulated without recognition of their ability to self-organize, their capacity to heal and repair breaks down, and they need increasing inputs and controls to be maintained.

Ecological and Socioeconomic Implications of Genetic Engineering

Genetic engineering has epistemological and ethical implications for the material conditions of our life, for our health and for our environment. Health implications are built into the very techniques of genetic engineering.

Genetic engineering moves genes across species by using 'vectors'—usually a mosaic recombination of natural genetic parasites from different sources, including viruses causing cancers and other diseases in animals and plants that are tagged with one or more antibiotic resistant 'marker' genes. Evidence accumulating over the past few years confirms the fear that these vectors constitute major sources of genetic pollution with drastic ecological and public health consequences. Vector-mediated horizontal gene transfer and recombination are found to be involved in generating new pandemic strains of bacterial pathogens.

Genetic engineering also has major ecological impacts, even though the biotechnology industry and regulatory agencies keep claiming that there have been no adverse consequences from the over 500 field releases in the United States.[23] Existing field tests are not designed to collect environmental data, and test conditions do not approximate production conditions that include commercial scale, varying environments, and time periods. Yet, as Phil J. Regal has stated, "this sort of non-data on non-releases has been cited in policy circles as though 500 true releases have now informed scientists that there are no legitimate scientific concerns."[24]

Two studies of detailed environmental impact assessment have verified the hazards posed by large-scale introduction of genetically engineered organisms in the field of agriculture.

At the 1994 annual meeting of the Ecological Society of America, researchers from Oregon State University reported on tests to evaluate a genetically engineered bacterium designed to convert crop waste into ethanol.

A typical root zone-inhabiting bacterium, *Klebsiella planticola*, was engineered with the novel ability to produce ethanol, and the engineered bacterium was added to enclosed soil chambers in which a wheat plant was growing. In one soil type, all the plants in soil with the engineered bacterium died, while plants in untreated soil remained healthy.

In all cases, mycorrhizal fungi in the root system were reduced by more than half, which ruined nutrient uptake and plant growth. This result was unpredicted. Reduction in this vital fungus is known to result in plants that are less competitive with weeds or more susceptible to disease. In low organic matter sandy soil, the plants died from ethanol produced by the engineered bacterium in the root system, while in high organic matter sandy or clay soil, changes in nematode density and species composition resulted in significantly decreased plant growth. The lead researcher, Dr. Elaine Ingham, concluded that these results imply that there can be significant and serious effects resulting from the addition of a genetically engineered micro-organism (GEM) to soil. The tests, using a new and comprehensive system, disproved earlier suggestions that there were no significant ecological effects.[25]

In 1994, research scientists in Denmark reported strong evidence that an oilseed rape plant genetically engineered to be herbicide tolerant transmitted its transgene to a weedy natural relative, *Brassica campestris* ssp. *campestris*. This transfer can take place in just two generations of the plant.

In Denmark, *B. campestris* is a common weed in cultivated oilseed rape fields, where selective elimination by herbicides is now impossible. The wild relative of this weed is spread over large parts of the world. One way to assess the risk of releasing transgenic oilseed rape is to measure the rate of natural hybridization with *B. campestris*, because certain transgenes could make its wild relative a more aggressive weed, and even harder to control.

Although crosses with *B. campestris* have been used in the breeding of oilseed rape, natural interspecific crosses with oilseed rape were generally thought to be rare. Artificial crosses by hand pollination carried out in a risk assessment project in the United Kingdom were reported to be unsuccessful. A few studies, however, have reported spontaneous hybridization between oilseed rape and the parented species *B. campestris* in field experiments. As early as 1962, hybridization rates of 0.3 to 88 percent were measured for oilseed rape and wild *B. campestris*. The results of the Danish team showed that high levels of hybridization can occur in the field. Their field tests revealed that between 9 and 93 percent of hybrid seeds were produced under different conditions.[26]

The transfer of herbicide resistance to wild, weedy relatives of crops threatens to create 'superweeds' that are resistant to herbicides and hence uncontrollable. As a strategy for Monsanto to sell more Round-Up, and Ciba Geigy to sell more Basta, genetically engineered herbicide-resistant crops make sense. Yet this strategy runs counter to a policy of sustainable agriculture, since it undermines the very possibility of weed control.

Just as the strategy of using genetic engineering for herbicide resistance fails to control weeds and instead carries the risk of creating 'super weeds', the strategy of genetically engineered crops for pest resistance fails to control pests and instead carries the risk of creating 'super pests'.

In 1996, nearly two million acres in the United States were planted with a genetically engineered cotton variety from Monsanto called 'Bollgard'. Monsanto's Bollgard cotton is a transgenic variety that has been engineered with DNA from the soil microbe *Bacillus thurengesis* (Bt) to produce proteins poisonous to the bollworm, a cotton pest. Monsanto charged the farmers a 'technology fee' of $79 per hectare in addition to the price of seed for 'peace of mind' through 'seasonlong plant control . . . that stops worm problems before they start'. The company collected $51 million in one year alone from this 'technology fee'.[27]

The technology, however, has already failed the farmers.

The bollworm infestation on the genetically engineered crop was more than 20 to 50 times the level that typically triggers spraying. Further, since Bt has been an important natural biological control agent used by organic farmers, the genetic engineering strategy undermines the organic strategy.

Besides the 'technology fee', Monsanto has also placed highly restrictive rules on farmers. As the company states:

> Monsanto is only licensing growers to use seed containing the patented Bollgard gene for one crop. Saving or selling the seed for replanting will violate the limited license and infringe upon the patent rights of Monsanto. This may subject you to prosecution under federal law.[29]

Monsanto 'owns' the crop when it comes to reaping millions of dollars in rent from farmers, but it does not own the costs or take responsibility for the hazards that its transgenic crop generates.

IPR monopolies are justified on grounds that corporations are given IPRs by society so that society can benefit from their contributions. The failure of the transgenic cotton shows that the assumption that IPRs will 'improve' agriculture does not always hold. Instead, what we have is an example of social and ecological costs generated for society in general and farmers in particular. IPRs on crop varieties that are creating ecological havoc is an unjust system of total privatization of benefits and total socialization of costs.

Monopolies linked to this unaccountable and unjust system prevent the development of ecologically sound and socially just practices. Further, they force an agricultural system on people that threatens the environment and human health.

Ironically, the imposition of monopolies and of genetically engineered products is at the core of the 'free trade' system. Legally, it is a free trade treaty, the Uruguay Round of GATT, that is forcing all countries to have IPRs in agriculture. Economically, the introduction of genetically engineered products is being forced on unwilling citizens and countries on the grounds of 'free trade', which, as the case of Monsanto's soya beans illustrates, translates into the absolute freedom of transnational corporations to force hazardous products on people.

World Food Day on October 16, 1996 was celebrated by 500 organizations from seventy-five countries calling for an international boycott of genetically engineered soya beans resistant to the chemical herbicide glyphaosate, which Monsanto sells as Round-Up. Monsanto had genetically engineered the soya bean to increase its herbicide sale.[30]

This was also a major controversy at the World Food Summit held in Rome in November 1996. Monsanto claimed its soya bean was distinctive and novel to get a patent, yet now says that the new soya bean is indistinguishable from the conventional bean—in order to mix the two types of soya beans together offshore and import them to European markets. Citizens are demanding that the genetically engineered soya be labelled under their 'right to know' and 'right to choose'.

Both the soya bean and cotton are now Monsanto monopolies since it acquired Agracetus (which has broad species patents for all transgenic cotton and soya) in May 1996 for $150 million. These patents are given on the basis of novelty, but that novelty is denied in the face of consumer resistance and concern over the safety of genetically engineered products.

As a technique, genetic engineering is very sophisticated. But as a technology for using biodiversity sustainably to meet human needs, it is clumsy. Transgenic crops reduce biodiversity by displacing diverse crops, which provide diverse sources of nutrition.

In addition, new health risks are being introduced through transgenic crops. Genetically engineered foods have the potential of introducing new allergies. They also carry the risk of 'biological pollution', of new vulnerability to disease, of one species becoming dominant in an ecosystem, and of gene transfer from one species to another.

In an experiment carried out in the United Kingdom by Dr. James Bishop, scorpion genes were introduced into a virus to make an insecticidal spray to kill caterpillars. The transgenic virus is assumed to be safe on grounds that it will not cross species boundaries for its target, even though there are plenty of examples of viruses and disease organisms finding new target species. Scientific evidence also shows that genetic engineering can create 'super viruses', viruses resistant to pesticides. Complacency on biosafety issues is therefore not justified on the basis of available scientific evidence.

A clearance has recently been given for the first trial of genetically engineered crops in India. They include a tomato engineered with Bt and hybrid brassica. There is already enough scientific evidence that genetic engineering with Bt is contributing to resistance, and therefore is not a sustainable route for controlling plant pests and disease.

The promised benefits of genetically engineered crops and foods are illusory, although their potential risks are real. The illusion of genetic engineering is, however, not merely at the systems level in food production and consumption. It is also at the scientific level. Genetic engineering offers its promises on the basis of genetic reductionism and determinism. Yet, both of these assumptions are being proved false through molecular biology research itself.

Celebrating and Conserving Life

In the era of genetic engineering and patents, life itself is being colonized. Ecological action in the biotechnology era involves keeping the self-organization of living systems free—free of technological manipulations that destroy the self-healing and self-organizational capacity of organisms, and free of legal manipulations that destroy the capacities of communities to search for their own solutions to human problems from the richness of the biodiversity that we have been endowed with.

There are two strands in my current work that respond to

the manipulation and monopolization of life. Through *Navdanya*, a national network in India for setting up community seed banks to protect indigenous seed diversity, we have tried to build an alternative to the engineering view of life. Through work to protect the intellectual commons—either in the form of Seed *Satyagraha* launched by the farmers' movement or in the form of the movement for common intellectual rights that we have launched with the Third World Network—we have tried to build an alternative to the paradigm of knowledge and life itself as private property.

It is this freedom of life and freedom to live that I increasingly see as the core element of the ecology movement as we reach the end of the millennium. And in this struggle, I frequently draw inspiration from the Palestinian poem 'The Seed Keepers':

Burn our land
burn our dreams
pour acid onto our songs
cover with sawdust
the blood of our massacred people
muffle with your technology
the screams of all that is free,
wild and indigenous.
Destroy
Destroy our grass and soil
raze to the ground
every farm and every village
our ancestors had built
every tree, every home
every book, every law
and all the equity and harmony.
Flatten with your bombs
every valley; erase with your edits
our past,
our literature; our metaphor

Denude the forests
and the earth
till no insect,
no bird
no word
can find a place to hide.
Do that and more.
I do not fear your tyranny
I do not despair ever
for I guard one seed
a little live seed
that I shall safeguard
and plant again.

Chapter Three

The Seed
and the Earth

Regeneration lies at the heart of life: it has been the central principle guiding sustainable societies. Without regeneration, there can be no sustainability. Modern industrial society, however, has no time for thinking about regeneration, and therefore no space for living regeneratively. Its devaluation of the proccsses of regeneration are the cause of both the ecological crisis and the crisis of sustainability.

In the *Rig Veda*, the hymn to the healing plants, medicinal plants are referred to as mothers because they sustain us.

> Mothers, you have a hundred forms
> and a thousand growths.
> You who have a hundred ways of
> working, make this person
> whole for me.
> Be joyful, you plants that bear
> flowers and those that bear fruit.

The continuity between regeneration in human and non-human nature that was the basis of all ancient worldviews was broken by patriarchy. People were separated from nature, and the creativity involved in processes of regeneration was denied. Creativity became the monopoly of men, who were considered to be engaged in production; women were engaged in mere reproduction or recreation which, rather than being treated as renewable production, was looked upon as non-productive.

The notion of activity being purely male was constructed

on the separation of the earth from the seed, and on the association of an inert and empty earth with the passivity of the
female. The symbols of the seed and the earth, therefore,
undergo a metamorphosis when cast in a patriarchal mould;
gender relations as well as our perception of nature and its
regeneration are also restructured. This unecological view of
nature and culture has formed the basis of patriarchal percep-
tions of gender roles in reproduction across religions and
through the ages.

This gendered seed/earth metaphor has been applied to
human production and reproduction to make the relationship
of dominance of men over women appear natural. But the
naturalness of this hierarchy is built on a material/spiritual
dualism, with male characteristics artificially associated with
pure spirit and female attributes constructed as merely material, bereft of spirit. As Johann Jacob Bachofen has stated:

> The triumph of paternity brings with it the liberation of the
> spirit from the manifestations of nature, a sublimation of human
> existence over the laws of material life. Maternity pertains to the
> physical side of man, the only thing he shares with animals; the
> paternal spiritual principle belongs to him alone. Triumphant
> paternity partakes of the heavenly light, while child-bearing
> motherhood is bound up with the earth that bears all things.[1]

Central to the patriarchal assumption of men's superiority
over women is the social construct of passivity/materiality as
female and animal, and activity/spirituality as male and dis-
tinctly human. This is reflected in dualisms like mind/body,
with the mind being non-material, male, and active, and the
body physical, female, and passive. It is also reflected in the
dualism of culture/nature, with the assumption that men alone
have access to culture while women are bound up with the
earth that bears all things.[2] What these artificial dichotomies
obscure is that activity, not passivity, is nature's forte.

The new biotechnologies reproduce the old patriarchal divi-
sions of activity/passivity, culture/nature. These dichotomies are

then used as instruments of capitalist patriarchy to colonize the regeneration of plants and human beings. Only by decolonizing regeneration can the activity and creativity of women and nature in a non-patriarchal mould be reclaimed.

Organisms: the New Colonies

The land, the forests, the rivers, the oceans, and the atmosphere have all been colonized, eroded, and polluted. Capital now has to look for new colonies to invade and exploit for its further accumulation—the interior spaces of the bodies of women, plants, and animals.

The invasion and takeover of land as colonies was made possible through the technology of the gunboat; the invasion and takeover of the life of organisms as the new colonies is being made possible through the technology of genetic engineering.

Biotechnology, as the handmaiden of capital in the post-industrial era, makes it possible to colonize and control that which is autonomous, free, and self-regenerative. Through reductionist science, capital goes where it has never been before. The fragmentation of reductionism opens up areas for exploitation and invasion. Technological development under capitalist patriarchy proceeds steadily from what it has already transformed and used up, driven by its predatory appetite, toward that which has still not been consumed. It is in this sense that the seed and women's bodies as sites of regenerative power are, in the eyes of capitalist patriarchy, among the last colonies.[3]

While ancient patriarchy used the symbol of the active seed and the passive earth, capitalist patriarchy, through the new biotechnologies, reconstitutes the seed as passive, and locates activity and creativity in the engineering mind. Five hundred years ago, when land began to be colonized, the reconstitution of the earth from a living system into mere matter went hand in hand with the devaluation of the contributions of non-European cultures and nature. Now, the reconstitution of the seed from a regenerative source of life into valueless raw material goes hand in hand with the devaluation of those who regenerate life through the seed—that is, the farmers and peasants of the Third World.

From *Terra Mater* to *Terra Nullius*

All sustainable cultures, in their diversity, have viewed the earth as *terra mater*. The patriarchal construct of the passivity of the earth and the consequent creation of the colonial category of land as *terra nullius* served two purposes: it denied the existence and prior rights of original inhabitants, and it negated the regenerative capacity and life processes of the earth.[4] The decimation of indigenous peoples everywhere was justified morally on the grounds that they were not really human; they were part of the fauna. As John Pilger has observed, the *Encyclopaedia Britannica* appeared to be in no doubt about this in the context of Australia: "Man in Australia is an animal of prey. More ferocious than the lynx, the leopard, or the hyena, he devours his own people."[5] In an Australian textbook, *Triumph in the Tropics*, Australian aborigines were equated with their half-wild dogs.[6] Being animals, the original Australians and Americans, Africans and Asians possessed no rights as human beings. Their lands could be usurped as *terra nullius*— lands empty of people, vacant, wasted, and unused. The morality of the missions justified the military takeover of resources all over the world to serve imperial markets. European men were thus able to describe their invasions as discoveries, their piracy and theft as trade, and their extermination and enslavement as a civilizing mission.

Scientific missions colluded with religious missions to deny rights to nature. The rise of mechanical philosophy with the emergence of the scientific revolution was based on the destruction of concepts of a self-regenerative, self-organizing nature, which sustained all life. For Francis Bacon, who is called the father of modern science, nature was no longer a mother, but rather a female, to be conquered by an aggressive masculine mind. As Carolyn Merchant points out, this transformation of nature from a living, nurturing mother to inert, dead, and manipulable matter was eminently suited to the exploitation imperative of growing capitalism. The nurturing earth image acted as a cultural constraint on the exploitation of nature. "One does not readily slay a mother, dig her entrails, or mutilate her body", writes Merchant. But the images of mastery and domination created by the Baconian programme and the scientific revolution removed all restraint, and functioned as cultural sanctions for the denudation of nature.

The removal of animistic, organic assumptions about the cosmos constituted the death of nature—the most far-reaching effect of the scientific revolution. Because nature was now viewed as a system of dead, inert particles moved by external (rather than inherent) forces, the mechanical framework itself could legitimize the manipulation of nature. Moreover, as a conceptual framework, the mechanical order was associated with a framework of values based on power, fully compatible with the directions taken by commercial capitalism.[7]

The construct of the inert earth was given a new and sinister significance, as development denied the earth's productive capacity and created systems of agriculture that could not regenerate or sustain themselves.

Sustainable agriculture is based on the recycling of soil nutrients. This involves returning to the soil part of the nutrients that come from it and support plant growth. The maintenance of the nutrient cycle, and through it the fertility of the soil, is based on an inviolable law of return that recognizes the earth as the source of fertility. The Green Revolution paradigm of agriculture substituted the regenerative nutrient cycle with linear flows of purchased inputs of chemical fertilizers from factories and marketed outputs of agricultural commodities. Fertility was no longer the property of soil, but of chemicals. The Green Revolution was essentially based on miracle seeds that needed chemical fertilizers and did not produce plant outputs for returning to the soil.[8] The earth was again viewed as an empty vessel, this time for holding intensive inputs of irrigated water and chemical fertilizers. The activity lay in the miracle seeds, which transcended nature's fertility cycles.

Ecologically, however, the earth and soil were not empty, and the growth of Green Revolution varieties did not take place only with the seed fertilizer packet. The creation of soil diseases and micronutrient deficiencies was an indication of the invisible demands the new varieties were making on the fertility of the soil; desertification indicated the broken cycles of soil fertility caused by an agriculture that produced only for the market. The increase in production of grain for marketing was achieved in the Green Revolution strategy by reducing the biomass for internal use on the farm. The reduction of output for straw production was probably not considered a serious cost, since chemical fertilizers were thought to be a total substitute for organic manure. Yet, as experience has shown, the fertility of soils cannot be reduced to nitrogen, phosphorus, and potassium in factories, and agricultural productivity necessarily includes returning to the soil part of the biological products that the soil yields. The seed and the earth mutually create conditions for each other's regeneration and renewal. Technologies cannot provide a substitute for nature and cannot work outside of nature's ecological processes without destroying the very basis of production, nor can markets provide the only measure of output and yield.

Biological products which were not sold on the market but used as inputs for maintaining soil fertility, were totally ignored by the cost-benefit equations of the Green Revolution miracle. They did not appear in the list of inputs because they were not purchased, nor in the list of outputs because they were not sold. Yet what was seen as unproductive or as waste in the commercial context of the Green Revolution is now emerging as productive in the ecological context, and as the only route to sustainable agriculture. By treating essential organic inputs as waste, the Green Revolution strategy unwittingly ensured that fertile and productive soils were actually laid waste; the land-augmenting technology has proved to be a land-degrading and land-destroying one. With the greenhouse effect and global warming, a new dimension has been added to the ecologically destructive effect of chemical fertilizers; nitrogen-based fertilizers release nitrous oxide, one of the greenhouse gases causing global warming, into the atmosphere. Chemical fertilizers have thus contributed to the erosion of food security through the pollution of land, water, and the atmosphere.

Seeds from the Lab

While the Green Revolution was based on the assumption that the earth is inert, the biotechnology revolution robs the seed of its fertility and self-regenerative capabilities, colonizing it in two major ways: through technical means and through property rights.

Processes like hybridization are the technological means that stop seed from reproducing itself. This provides capital with an eminently effective way of circumventing natural constraints on the commodification of the seed. Hybrid varieties do not produce true-to-type seed, and farmers must return to the breeder each year for new seed stock.

To use Jack Kloppenburg's description of the seed: it is both a means of production and a product.[9] Whether they are tribespeople engaged in shifting cultivation or peasants practicing settled agriculture, in planting each year's crop, farmers also reproduce the necessary element of their means of production. The seed thus presents capital with a simple biological obstacle: given the appropriate conditions, it reproduces itself and multiplies. Modern plant breeding has primarily been an attempt to remove this biological obstacle, and the new biotechnologies are the latest tools for transforming what is simultaneously a means of production and a product into mere raw material.

The hybridization of seed was an invasion into the seed itself. As E. Goppenburg has stated, it broke the unity of the seed as food grain and as a means of production. In doing so, it opened up the space for capital accumulation that private industry needed in order to control plant breeding and commercial seed production. And it became the source of ecological disruption by transforming a self-regenerative process into a broken linear flow of supply of living seed as raw material, and a reverse flow of seed commodities as products. The decoupling of seed from grain also changes the status of seed.

The commodified seed is ecologically incomplete and ruptured at two levels: First, it does not reproduce itself, whereas by definition seed is a regenerative resource. Genetic resources are thus, through technology, transformed from a renewable into a non-renewable resource. Second, it does not produce by itself; it needs the help of other purchased inputs. And, as the seed and chemical companies merge, the dependence of inputs will increase. Whether a chemical is added externally or internally, it remains an external input in the ecological cycle of the reproduction of seed. It is this shift from ecological processes of production through regeneration to technological processes of non-regenerative production that underlies the dispossession of farmers and the drastic reduction of biological diversity in agriculture. It is at the root of the creation of poverty and of nonsustainability in agriculture.

Where technological means fail to prevent farmers from reproducing their own seed, legal regulations in the forms of intellectual property rights and patents are brought in. Patents are central to the colonization of plant regeneration and, like land titles, are based on the assumption of ownership and property. As the vice president of Genentech has stated: "When you have a chance to write a clean slate, you can make some very basic claims, because the standard you are compared to is the state of prior art, and in biotechnology there just is not much." [10] Ownership and property claims are made on living resources, but prior custody and use of those resources by farmers is not the measure against which the patent is set. Rather, it is the intervention of technology that determines the claim to their exclusive use. The possession of this technology, then, becomes the reason for ownership by corporations, and for the simultaneous dispossession and disenfranchisement of farmers.

As with the transformation of *terra mater* to *terra nullius*, the new biotechnologies rob farmers' seeds of life and value by the very process that makes corporate seeds the basis of wealth creation. The production and use of landraces (indigenous varieties evolved through both natural and human selection) are essential to Third World farmers. These are termed 'primitive' cultivars, whereas those varieties created by modern plant breeders in international research centres or by transnational seed corporations are called 'advanced' or 'elite'. Trevor Williams, the former Executive Secretary of the International Board for Plant Genetic Resources, has argued that it is not the original material that produces cash returns, and at a 1983 forum on plant breeding, stated that raw germ plasm only becomes valuable after considerable investment of time and money.[11] According to this calculation, peasants' time is considered valueless and available for free. Once again, all prior processes of creation are being denied and devalued by defining them as nature. Thus, plant breeding by farmers is not breeding; real breeding is seen to begin when this 'primitive germ plasm' is mixed or crossed with inbred lines in international labs by international scientists. That is, innovation occurs only through the long, laborious, expensive, and always risky process of backcrossing and other means required to first make genetic sense out of the chaos created by the foreign germ plasm, and eventually to make dollars and cents from a marketable product.[12]

But the landraces that farmers have developed are not genetically chaotic. Nor do they lack innovation. They consist of improved and selected material, embodying the experience, inventiveness and hard work of farmers past and present; the evolutionary material processes they have undergone serve ecological and social needs. These needs are now being under- mined by the monopolizing tendency of corporations. Placing the contributions of corporate scientists over and above the intellectual contributions made by Third World farmers over 10,000 years—contributions to conservation, breeding, domestication, and development of plant and animal genetic resources—is based on rank social discrimination.

IPRs vs. Farmers' and Plant Breeders' Rights

As Pat Mooney has argued, "The perception that intellectual property is only recognizable when produced in laboratories by men in lab coats is fundamentally a racist view of scientific development." [13] Indeed, the total genetic change achieved by farmers over millennia has been far greater than that achieved during the last 100 to 200 years of more systematic science- based efforts. The limits of the market system in assigning value can hardly be a reason for denying value to farmers' seeds and nature's seeds. It indicates the deficiencies in the logic of the market rather than the status of the seed or of farmers' intelligence.

The denial of prior rights and creativity is essential for owning life. A brief book prepared by the biotechnology industry states:

Patent laws would in effect have drawn an imaginary line around your processes and products. If anyone steps over that line to use, make or sell your inventions or even if someone steps over that line in using, making or selling his own products, you could sue for patent protection. [14]

Jack Doyle has appropriately remarked that patents are less concerned with innovation than with territory, and can act as instruments of territorial takeover by claiming exclusive access to creativity and innovation, thereby monopolizing rights to ownership. [15] The farmers, who are the guardians of the germ plasm, have to be dispossessed to allow the new colonization to happen.

As with the colonization of land, the colonization of life processes will have a serious impact on Third World agriculture. First, it will undermine the cultural and ethical fabric of agriculturally based societies. For instance, with the introduction of patents, seeds—which have hitherto been treated as gifts and exchanged freely between farmers—will become patented commodities. Hans Leenders, former Secretary General of the International Association of Plant Breeders for the Protection of Plant Varieties, has proposed the abolition of the farmer's right to save seed. He says:

> Even though it has been a tradition in most countries that a farmer can save seed from his own crop, it is under the changing circumstances not equitable that farmers can use this seed and grow a commercial crop out of it without payment of a royalty; the seed industry will have to fight hard for a better kind of protection.[16]

Although genetic engineering and biotechnology only relocate existing genes rather than create new ones, the ability to relocate and separate is translated into the power and right to own. The power to own a part is then translated into control of the entire organism.

Additionally, the corporate demand for the conversion of a common heritage into a commodity, and for profits generated through this transformation to be treated as property rights, will have serious political and economic implications for Third World farmers. They will now be forced into a three-level relationship with the corporations demanding a monopoly on life forms and life processes through patents. First, farmers are suppliers of germ plasm to transnational corporations; second, they become competitors in terms of innovation and rights to genetic resources; and third, they are consumers of the technological and industrial products of these corporations. In other words, patent protection transforms farmers into suppliers of free raw material, displaces them as competitors, and makes them totally dependent on industrial supplies for vital inputs such as seed. The frantic cry for patent protection in agriculture is really a ruse for control of biological resources in agriculture. It is argued that patent protection is essential for innovation, yet it is essential only for that innovation that garners profit for corporate business. After all, farmers have been making innovations for centuries, as have public institutions for decades, without property rights or patent protection.

Further, unlike plant breeders' rights (PBRs), the new utility patents are very broad-based, allowing monopoly rights over individual genes and even characteristics. PBRs do not entail ownership of the germ plasm in the seeds, they only grant a monopoly right over the selling and marketing of a specific variety. Patents, on the other hand, allow for multiple claims that may cover not only whole plants, but plant parts and processes as well. So, according to attorney Anthony Diepenbrock:

> You could file for protection of a few varieties of crops, their macroparts (flowers, fruits, seeds and so on), their microparts (cells, genes, plasmids and the like) and whatever novel processes you develop to work these parts, all using one multiple claim.[17]

Patent protection implies the exclusion of farmers' rights over resources having these genes and characteristics, undermining the very foundation of agriculture. For example, a patent has been granted in the United States to a biotechnology company, Sungene, for a sunflower variety with very high oleic acid content. The claim allowed was for the characteristic (i.e. high oleic acid) and not just for the genes producing the characteristic. Sungene has notified sunflower breeders that the development of any variety high in oleic acid will be considered an infringement of its patent.

The landmark event for the patenting of plants was the 1985 judgment in the United States, now famous as *ex parte*

Hibberd, in which molecular genetics scientist Kenneth Hibberd and his co-inventors were granted patents on the tissue culture, seed, and whole plant of a maize line selected from tissue culture.[18] The Hibberd application included over 260 separate claims, which gave the molecular genetics scientists the right to exclude others from use of all 260 aspects. While Hibberd apparently provides a new legal context for corporate competition, the most profound impact will be felt in the competition between farmers and the seed industry.

As Kloppenburg has indicated, with the Hibberd judgement a juridical framework is now in place to allow the seed industry to realize one of its longest held and most cherished goals: to force all farmers to buy seed every year instead of obtaining it through reproduction. Industrial patents allow others to use a product, but deny them the right to make it. Since seed makes itself, a strong utility patent for seed implies that a farmer purchasing patented seed would have the right to use (to grow) the seed, but not to make it (to save and replant). If the Dunkel Draft of the GATT is implemented, the farmer who saves and replants the seed of a patented or protected plant variety will be violating the law.

Through intellectual property rights an attempt is made to take away what belongs to nature, to farmers and to women, and to term this invasion 'improvement' and 'progress'. Violence and plunder as instruments of wealth creation are essential to the colonization of nature and of our bodies through the new technologies. Those who are exploited become the criminals, those who exploit require protection. The North must be protected from the South so that it can continue its uninterrupted theft of the Third World's genetic diversity. The seed wars, trade wars, patent protection, and intellectual property rights in the GATT are claims to ownership through separation and fragmentation. If the regime of rights being demanded by the United States is implemented, the transfer of funds from poor to rich countries will exacerbate the Third World crisis 10 times over.[19]

The United States has accused the Third World of piracy. The estimates for royalties lost are $202 million per year for agricultural chemicals and $2.5 billion annually for pharmaceuticals.[20] In a 1986 U.S. Department of Commerce survey, U.S. companies claimed they lost $23.8 billion yearly due to inadequate or ineffective protection of intellectual property. Yet, as the team at the Rural Advancement Foundation International in Canada has shown, if the contributions of Third World peasants and tribespeople are taken into account, the roles are dramatically reversed: the United States would owe Third World countries $302 million in agriculture royalties and $5.1 billion for pharmaceuticals. In other words, in these two biological industry sectors alone, the United States should owe $2.7 billion to the Third World.[21] It is to prevent these debts from being taken into account that it becomes essential to set up the creation boundary through the regulation of intellectual property rights; without it, the colonization of the regenerative processes of life renewal is impossible. Yet if this, too, is allowed to happen in the name of patent protection, innovation and progress, life itself will have been colonized.

There are, at present, two trends reflecting different views as to how native seeds, indigenous knowledge, and farmers' rights should be treated. On the one hand, there are initiatives across the world that recognize the inherent value of seeds and biodiversity, acknowledge the contributions of farmers to agricultural innovation and seed conservation, and see patents as a threat both to genetic diversity and to farmers. At the global level, the most significant platforms to have made the issue of farmers' rights visible are the Food and Agriculture Organization (FAO) Commission of Plant Genetic Resources[22] and the Keystone Dialogue.[23] At the local level, communities all over Asia, Africa, and Latin America are taking steps to save and regenerate their native seeds. As mentioned in Chapter Two, we have set up a network in India called *Navdanya*, meaning 'native seed conservation'.

Despite these initiatives, however, the dominant trend continues to be toward the displacement of local plant diversity and its substitution by patented varieties. At the same time, international agencies under pressure from seed corporations are pushing for regimes of intellectual property rights that deny farmers their intellect and their rights. The March 1991 revision of the International Convention for the Protection of New Varieties of Plants, for example, allows countries to remove the farmers' exemption—the right to save and replant seed—at their discretion.[24]

In another development leading to the privatization of genetic resources, the Consultative Group on International Agricultural Research made a policy statement on May 22, 1992 allowing for the privatization and patenting of genetic resources held in international gene banks.[25] The strongest pressure for patents is coming from the GATT, especially in relation to the agreement on TRIPs and agriculture.[26]

Engineering Humans

Just as technology changes seed from a living, renewable resource into mere raw material, it devalues women in a similar way. For instance, reproduction has been linked to the mechanization of the female body, in which a set of fragmented fetishized and replaceable parts are managed by professional medical experts. While most advanced in the United States, it is also spreading to the Third World.

The mechanization of childbirth is evident in the increased use of Caesarean sections. Significantly, this method, which requires the most management by the doctor and the least labour by the woman, is seen as providing the best product. But Caesarean sections are a surgical procedure, and the chances of complications are two to four times greater than during normal vaginal delivery. They were introduced as a means of delivering babies at risk, but when they are done routinely, they can pose an unnecessary threat to health and even life. Almost one in every four Americans is now born by Caesarean section.[27] Brazil has one of the highest proportions of Caesarean section deliveries in the world; a nationwide study of patients enrolled in the social security system showed an increase in the proportion of Caesarean sections from 15 percent in 1974 to 31 percent in 1980. In urban areas, such as the city of São Paulo, rates as high as 75 percent have been observed.

As with plant regeneration, where agriculture has moved from the Green Revolution technologies to biotechnology, a parallel shift is also taking place with regard to human reproduction. With the introduction of new reproductive technologies, the relocation of knowledge and skills from the mother to the doctor, from women to men, will be accentuated. Peter Singer and Deane Wells, in *The Reproductive Revolution*, have suggested that the production of sperm is worth a great deal more than the production of eggs. They conclude that sperm vending places a greater strain on the man than egg donation does on the woman, in spite of the chemical and mechanical invasion of her body.[28]

Whereas *in vitro* fertilization and other technologies are currently only offered for abnormal cases of infertility, the boundary between nature and non-nature is fluid, and normality has a tendency to be redefined as abnormality when technologies created for abnormal cases become more widely used. When pregnancy was first transformed into a medical disease, professional management was limited to abnormal cases, while normal cases continued to be looked after by the original professional, the midwife. While 70 percent of childbirths in the United Kingdom were thought normal enough to be delivered at home in the 1930s, by the 1950s the same percentage were identified as abnormal enough to be delivered in the hospital!

The new reproductive technologies have provided contemporary scientific rhetoric for the reassertion of an enduring set of deeply patriarchal beliefs. The idea of women as vessels, and the foetus as created by the father's seed and owned by patriarchal right, leads logically to the breaking of organic links between the mother and the foetus.

Medical specialists, falsely believing that they produce and create babies, force their knowledge on knowing mothers. They treat their own knowledge as infallible, and women's knowledge as wild hysteria. And through their fragmented and invasive knowledge, they create a maternal-foetal conflict in which life is seen only in the foetus, and the mother is reduced to a potential criminal threatening her baby's life.

The false construction of a maternal-foetal conflict, which was the basis of the patriarchal takeover of childbirth by male medical practitioners from women and midwives, was adopted by feminists as the basis of women's 'choice' a century later. The 'pro-choice' and 'pro-life' movements are thus both based on a patriarchal construction of women and reproduction.

The medical construction of life through technology is often inconsistent with the living experience of women as thinking and knowing human beings. When such conflicts arise, patriarchal science and law have worked hand in hand to establish the control by professional men over women's lives, as demonstrated by recent work on surrogacy and the new reproductive technologies. Women's rights, linked with their regenerative capacities, have been replaced by those of doctors as producers, and rich, infertile couples as consumers.

The woman whose body is being exploited as a machine is not seen as the one who needs protection from doctors and rich couples. Instead, the consumer, the adoptive male parent, needs protection from the biological mother, who has been reduced to a surrogate uterus. This is exemplified in the famous 1986 Baby M. case, in which Mary Beth agreed to lend her uterus, but after experiencing what having a baby meant, wanted to return the money and keep the child. A New Jersey judge ruled that a man's contract with a woman concerning his sperm is sacred, and that pregnancy and childbirth are not. Commenting on this notion of justice, in her book *Sacred Bond*, Phylis Chesler says: "It's as if these experts were 19th century missionaries and Mary Beth a particularly stubborn native who refused to convert to civilization, and what's more, refused to let them plunder her natural resources without a fight." [29]

The role of man as creator has also been taken to absurd lengths in an application submitted for a patent for the characterization of the gene sequence coding for human relaxin, a hormone that is synthesized and stored in female ovaries and helps in dilation, thus facilitating the birth process. A naturally occurring substance in women's bodies is being treated as the invention of three male scientists: Peter John Hud, Hugh David Nill, and Geoffrey William Tregear. [30] Ownership is thus acquired through invasive and fragmenting technology, and it is this link between fragmenting technology and control and ownership of resources and people that forms the basis of the patriarchal project of knowledge as power over others.

Such a project is based on the acceptance of three separations: the separation of mind and body; the gendered separation of male activity as intellectual and female activity as biological; and the separation of the knower and the known. These separations allow for the political construction of a creation boundary that divides the thinking, active male from the unthinking, passive female, and from nature.

Biotechnology is today's dominant cultural instrument for carving out the boundary between nature and culture through intellectual property rights, and for defining women's and farmers' knowledge and work as nature. These patriarchal constructs are projected as natural, although there is nothing natural about them. As Claudia Von Werlhof has pointed out, from the dominant standpoint, nature is everything that should be available free or as cheaply as possible. This includes the products of social labour. The labour of women and Third World farmers is said to be non-labour, mere biology, a natural resource; their products are thus akin to natural deposits. [31]

The Production and Creation Boundaries

The transformation of value into disvalue, labour into non-labour, knowledge into non-knowledge, is achieved by two very powerful constructs: the production boundary and the creation boundary.

The production boundary is a political construct that excludes regenerative, renewable production cycles from the domain of production. National accounting systems, which are used for calculating growth through the gross national product, are based on the assumption that if producers consume what they produce, they do not, in fact, produce at all because they fall outside the production boundary.[32] All women who produce for their families, children, and nature are thus treated as unproductive, as economically inactive. Discussions at the United Nations Conference on Environment and Development on issues of biodiversity have also referred to production for one's own consumption as a market failure.[33] When economies are confined to the marketplace, self-sufficiency in the economic domain is seen as economic deficiency. The devaluation of women's work, and of work done in subsistence economies in the Third World, is the natural outcome of a production boundary constructed by capitalist patriarchy.

The creation boundary does to knowledge what the production boundary does to work: it excludes the creative contributions of women as well as Third World peasants and tribespeople, and views them as being engaged in unthinking, repetitive, biological processes. The separation of production from reproduction, the characterization of the former as economic and the latter as biological, are some of the underlying assumptions that are treated as natural even though they have been socially and politically constructed.

This patriarchal shift in the creation boundary is misplaced for many reasons. First, the assumption that male activity is true creation because it takes place *ex nihilo* is ecologically false. No technological artifact or industrial commodity is formed out of nothing; no industrial process takes place where nothing was before. Nature and its creativity as well as people's social labour are consumed at every level of industrial production as raw material or energy. The biotech seed that is treated as a creation to be protected by patents could not exist without the farmer's seed. The assumption that only industrial production is truly creative because it produces from nothing hides the ecological destruction that goes with it. The patriarchal creation boundary allows ecological destruction to be perceived as creation, and ecological regeneration as underlying the breakdown of ecological cycles and the crisis of sustainability. To sustain life means, above all, to regenerate life; but according to the patriarchal view, to regenerate is not to create, it is merely to repeat.

Such a definition of creativity is also false because it fails to see that women's and subsistence producers' work go into child-rearing and cultivation, both of which conserve regenerative capacity.

The assumption that creation implies the reduction of novelty is also false; regeneration is not merely repetition. It involves diversity, while engineering produces uniformity. Regeneration, in fact, is how diversity is produced and renewed. While no industrial process takes place out of nothing, the creation myth of patriarchy is particularly unfounded in the case of biotechnologies, where life forms are the raw material for industrial production.

Rebuilding Connections

The source of patriarchal power over women and nature lies in separation and fragmentation. Nature is separated from and subjugated to culture; mind is separated from and elevated above matter; female is separated from male, and identified with nature and matter. The domination over women and nature is one outcome; the disruption of cycles of regeneration is another. Disease and ecological destruction arise from this interruption of the cycles of renewal of life and health. The crises of health and ecology suggest that the assumption of man's ability to totally engineer the world, including seeds and women's bodies, is in question. Nature is not the essentialized, passive construct that patriarchy assumes it to be. Ecology forces us to recognize the disharmonies and harmonies in our interactions with nature. Understanding and sensing connections and relationships is the ecological imperative.

The main contribution of the ecology movement has been the awareness that there is no separation between mind and body, human and nature. Nature consists of the relationships and connections that provide the very conditions for our life and health. This politics of connection and regeneration provides an alternative to the politics of separation and fragmentation that is causing ecological breakdown. It is a politics of solidarity with nature. This implies a radical transformation of nature and culture in such a manner that they are mutually permeating, not separate and oppositional. By starting a partnership with nature in the politics of regeneration, women are simultaneously reclaiming their own and nature's activity and creativity. There is nothing essentialist about this politics because it is, in fact, based on denying the patriarchal definition of passivity as the essence of women and nature. There is nothing absolutist about it because the natural is constructed through diverse relationships in diverse settings. Natural agriculture and natural childbirth involve human creativity and sensitivity of the highest order, a creativity and knowledge emerging from partnership and participation, not separation. The politics of partnership with nature, as it is being shaped in the everyday lives of women and communities, is a politics of rebuilding connections, and of regeneration through dynamism and diversity.

Chapter Four

Biodiversity and People's Knowledge

The tropics are the cradle of the planet's biological diversity, with a multiplicity of ecosystems.[1] A majority of Third World countries are located in the tropics and are thus endowed with this wealth of biological diversity, which is being rapidly destroyed. The two primary causes for the large-scale destruction of this biodiversity are:

1. Habitat destruction due to internationally financed megaprojects—such as the building of dams, highways, mines, and aquaculture in areas rich in biological diversity. The Blue Revolution is an example of how coastal areas rich in marine diversity and inland areas rich in agricultural diversity are being destroyed through intensive shrimp farming.

2. The technological and economic push to replace diversity with homogeneity in forestry, agriculture, fishery, and animal husbandry. The Green Revolution is an example of the deliberate replacement of biological diversity with biological uniformity and monocultures.

Biodiversity erosion starts a chain reaction. The disappearance of one species is related to the extinction of innumerable other species with which it is interrelated through food webs and food chains. The crisis of biodiversity, however, is not just a crisis of the disappearance of species, which serve as industrial raw material and have the potential of spinning dollars for corporate enterprises. It is, more basically, a crisis that threatens the life support systems and livelihoods of millions of people in Third World countries.

Biodiversity is a people's resource. While the industrialized world and affluent societies turned their backs to biodiversity, the poor in the Third World have continued to depend on biological resources for food and nutrition, for health care, for energy, for fibre, and for housing.

The emergence of the new biotechnologies has changed the meaning and value of biodiversity. It has been converted from a life support base for poor communities to the raw material base for powerful corporations. Even though references are increasingly made to global biodiversity and global genetic resources, biodiversity—unlike the atmosphere or the oceans—is not a global commons in the ecological sense. Biodiversity exists in specific countries and is used by specific communities. It is global only in its emerging role as raw material for global corporations.

The emergence of new intellectual property regimes, and new and accelerated potential for exploitation of biodiversity, create new conflicts over biodiversity—between private and common ownership, between global and local use.

Biodiversity: Whose Resource?

Biodiversity has always been a local common resource. A resource is common property when social systems exist to use it on the principles of justice and sustainability. This involves a combination of rights and responsibilities among users, a combination of utilization and conservation, a sense of co-production with nature and of gift-giving among members of the community.

There are many levels at which resource ownership and the concept of knowledge and access to it differs between private property regimes and common property systems. Common property systems recognize the intrinsic worth of biodiversity; regimes governed by IPRs see value as created through commercial exploitation. Common property knowledge and

resource systems recognize creativity in nature. As the vision-
ary biologist John Todd has stated, biodiversity carries the
intelligence of three and a half billion years of experimentation by life forms. Human production is viewed as co-production and co-creativity with nature. IPR regimes, in
contrast, are based on the denial of creativity in nature. Yet they
usurp the creativity of emerging indigenous knowledge and
the intellectual commons. Further, since IPRs are more a protection of capital investment than a recognition of creativity
per se, there is a tendency for ownership of knowledge, and the
products and processes emerging from it, to move toward areas
of capital concentration and away from poor people without
capital. Knowledge and resources are, therefore, systematically
alienated from the original custodians and donors, becoming
the monopoly of the transnational corporations.

Through this trend, biodiversity is converted from a local
commons into an enclosed private property. Indeed, the
enclosure of the commons is the objective of IPRs in the
areas of life forms and biodiversity. This enclosure is being
universalized through the TRIPs treaty of the GATT and certain interpretations of the Biodiversity Convention. It is also
the underlying mechanism of bioprospecting contracts.

Central to the privatization of knowledge and biodiversity
is the devaluation of local knowledge, the displacement of local
rights, and simultaneously the creation of monopoly rights to
biodiversity utilization through the claim of novelty. It has
sometimes been argued that monopolies exist even in traditional communities. Yet in the case of agriculture, for example,
seeds and knowledge are freely exchanged as gifts. Similarly,
knowledge of medicinal plants is a local common resource.

Plant-based systems of healing fall into two categories—
folk systems, and specialized systems like Ayurveda, Siddha,
and Unani. Even specialized systems, however, depend on folk
knowledge. In the Ayurveda classic, *Charaka Samhita*, indigenous medical practitioners are advised:

By knowing from cowherds, tapasvis, forest dwellers, hunters, gardeners, and by knowing about their form and properties, learn about herbs and medicinal plants.[2]

Ayurvedic knowledge is also part of the everyday knowledge of people. Folk traditions and specialized medical systems support each other, unlike in the medical-industrial system dominated by pharmaceutical corporations, where people do not figure as knowing subjects.

Non-Western medical systems also differ from the medical-industrial system of the West in that indigenous medical practitioners do not exercise a commercial monopoly through their practice. While they might not exchange their knowledge freely, they do freely gift its benefits. They do not use their knowledge to amass limitless private profit and wealth. They practice what we in India call *gyan daan*—the gifting of knowledge.

By their very logic, on the other hand, IPRs exploit knowledge for profit by excluding others from its use during the life of the patent. Since IPRs are often based on local knowledge and on tinkering with biodiversity that has hitherto been in the commons, they amount to an intellectual and material enclosure. Consequently, people lose access to the knowledge and resources vital to their survival and creativity, and to the conservation of cultural and biological diversity.

Two important historical tendencies surround the issue of knowledge. On one side, there is a growing recognition that the Western paradigm of mechanistic reductionism is at the root of the ecological and health crises and that non-Western systems of knowledge are better adapted to respect life. On the flip side, precisely when indigenous systems of knowledge could come into their own, the GATT is using IPRs to reinforce the monopoly of Western systems and devalue indigenous systems, even while exploiting them in order to set up IPR monopolies.

Indigenous Knowledge and IPRs

The patenting of products and processes derived from plants on the basis of indigenous knowledge has become a major issue of conflict in the IPR domain. The patenting of neem is but one example.

Neem, *Azarichdita indica*, a beautiful tree native to India, has been used for centuries as a biopesticide and a medicine. In some parts of India, the new year begins with eating the tender shoots of the neem tree. In other parts, the neem tree is worshipped as sacred. Everywhere in India, people begin their day by using the neem *datun* (toothbrush) to protect their teeth with its medicinal and anti-bacterial properties. Communities have invested centuries of care, respect, and knowledge in propagating, protecting, and using neem in fields, field bunds, homesteads, and common lands.

Today, this heritage is being stolen under the guise of IPRs. For centuries, the Western world ignored the neem tree and its properties: the practices of Indian peasants and doctors were not deemed worthy of attention by the majority of British, French, and Portuguese colonists. In the last few years, however, growing opposition to chemical products in the West, in particular pesticides, has led to a sudden enthusiasm for the pharmaceutical properties of neem. Since 1985, over a dozen U.S. patents have been taken out by U.S. and Japanese firms on formulas for stable neem-based solutions and emulsions—and even for a neem-based toothpaste. At least four of these are owned by W. R. Grace of the United States, three by the Native Plant Institute, another U.S. company, and two by the Japanese Terumo Corporation. Having garnered their patents, and with the prospect of a licence from the U.S. Environmental Protection Agency (EPA), Grace has set about manufacturing and commercializing its products by establishing a base in India. The company approached several Indian manufacturers with proposals to buy up their technology or to convince them to stop producing value-added products

and instead supply Grace with raw material. Grace is likely to
be followed by other patent-holding companies. "Squeezing
bucks out of the neem ought to be relatively easy," observes
Science magazine.[3]

The journal *Ag Biotechnology News* has called W. R Grace's
processing plant the "world's first neem tree based biopesticide
facility". Nearly every home and village in India, however, has
biopesticide facilities. The Indian cottage industries' organization (Khadi) and the Village Industries Commission have been
using and selling neem products for 40 years. Private entrepre-
neurs, too, have launched neem pesticides, such as 'Indiara'.
Neem toothpaste has been manufactured for decades by
Calcutta Chemicals, an indigenous company. W. R Grace's justification for the patents hinges on the claim that their mod-
ernized extraction processes constitute a genuine innovation:

> Although traditional knowledge inspired the research and
> development that led to these patented compositions and
> processes, they were considered sufficiently novel and different
> from the original product of nature and the traditional method
> of use to be patentable.[4]

In short, the processes are supposedly novel, an advance on
Indian techniques. This novelty, however, exists mainly in the
context of the ignorance of the West. Over the 2,000 years
that neem-based biopesticides and medicines have been used
in India, many complex processes were developed to make
them available for specific use, though the active ingredients
were not given Latinized scientific names. Common knowledge and use of neem were the primary reasons given by the
Indian Central Insecticide Board for not registering neem
products under the Insecticides Act of 1968. The Board
argued that neem materials had been in extensive use in India
for various purposes since time immemorial, without any
known deleterious effects.[5]

Biodiversity has different properties that can be utilized for
meeting human needs. In the case of neem, the knowledge

that the tree has biopesticidal properties is metaknowledge—knowledge of principles—in the public domain. Given this knowledge, various processes of technology can be used for preparing a variety of products from neem. These are obvious, not novel.

At the level of microknowledge—knowledge involved in tinkering with technical processes—the basis of IPR claims to neem is illegitimate on two grounds. First, it claims nature's creativity and the creativity of other cultures as its own. Second, in the case of neem, this leads to the false claim that the biopesticide property was created by the patentee. It treats petty tinkering as a source of creation, rather than acknowledging that specific species are the source of the creation of specific properties and characteristics, and that communities are the source of the knowledge that allows that property to be utilized.

The issue of IPRs is closely related to the issue of value. If all value is seen as being associated with capital, tinkering becomes necessary to add value. Simultaneously, value is taken away from the source (biological resources as well as indigenous knowledge), which is reduced to raw material.

Tinkering, however, does not create value. The value of the product is dependent on the source—in this case, neem—not on how it is processed. The same tinkering applied to another species would not produce a pesticide. Society is the source of the knowledge that neem makes a biopesticide, not the inventor of epistemologically petty but technologically powerful tinkering.

IPRs allow for the privatization of biodiversity and the intellectual commons. 'Bioprospecting' is increasingly the word used to describe this new form of enclosure.

Bioprospecting vs. People's Knowledge

Biodiversity has been protected through the flourishing of cultural diversity. Utilizing indigenous knowledge systems, cultures have built decentralized economies and production systems that use and reproduce biodiversity. Monocultures, by contrast, which are produced and reproduced through centralized control, consume biodiversity.

The challenge of biodiversity conservation is to enlarge the scope of economies based on diversity and decentralization, and shrink the scope of economies based on monocultures, monopolies, and non-sustainability. While both kinds of economies use biodiversity as an input, only economies based on diversity produce diversity. Monoculture economies produce monocultures.

When indigenous systems of knowledge and production interact with dominant systems of knowledge and production, it is important to anticipate whether the future options of the indigenous system or the dominant system will grow. Whose knowledge and values will shape the future options of diverse communities?

The World Resources Institute has defined this bioprospecting as the exploration of commercially valuable genetic and biochemical resources.[6] The metaphor is borrowed from the prospecting for gold or oil. While biodiversity is fast becoming the green gold and green oil for the pharmaceutical and biotechnology industries, suggesting that the use and value of biodiversity lies with the prospector, it is actually held by local indigenous communities. Further, this metaphor suggests that prior to prospecting, the resource lies buried, unknown, unused, and without value. Unlike gold or oil deposits, however, the uses and value of biodiversity are known by the communities from where the knowledge is taken through bioprospecting contracts.

The metaphor of bioprospecting thus hides the prior use, knowledge, and rights associated with biodiversity. Alternative economic systems disappear, and the Western prospector is projected as the only source for medical and agricultural uses of biodiversity. With the disappearance of alternatives, monopolies in the form of intellectual property rights appear natural.

When alternative and freely exchanged knowledge—such

as the use of neem or medicinal plants—is eclipsed, corporations with IPR protection appear to be the only source of biological pesticides or, for example, the cure for cancer. Their exclusive claims to added value and monopoly rights to production are rendered legitimate in the absence of alternatives, which, even if kept alive, are recognized as illegitimate.

The bias that use and value are only generated by Western corporations is apparent in most Western analyses of bioprospecting. As one proponent states:

> As industry interest in genetic and biochemical resources increases and more research and conservation institutions realize that they must use or face losing their countries' biodiversity, contractual agreements between collectors and suppliers of biological samples, and pharmaceutical and biotech companies, will become more important. Through the relationships they represent, these contracts can ensure that a portion of the value generated from developing genetically or biologically derived products is captured by the country and people who have been biodiversity's custodians.[7]

The concept of adding value through bioprospecting hides the removal and destruction of the value of indigenous plants and knowledge. As the genes of a particular plant gain value, the plant itself becomes dispensable, especially if the genes can be replicated *in vitro*. As useful characteristics of plants are identified by indigenous communities, the communities themselves—along with their lifestyles and knowledge systems—become dispensable.

It is important to view bioprospecting in the context of markets for patent commodities in the agricultural as well as the health sectors. The same corporations that prospect for the commercialization of biodiversity also displace economies based on alternative values and knowledge systems in order to expand their markets for seeds, biopesticides, and pharmaceuticals.

When indigenous communities are asked to sell their

knowledge to corporations, they are being asked to sell their birthright to continue to practice their traditions in the future, and to provide for themselves through their knowledge and their resources. This has already happened in the case of seeds in the industrialized world and in the case of plant-based medicines derived from Third World knowledge. Of the 120 active compounds currently isolated from the higher plants and widely used in modern medicine, 75 percent have uses that were known in traditional systems. Fewer than a dozen are synthesized by simple chemical modification; the rest are extracted directly from plants and then purified.[8] The use of traditional knowledge reportedly increases the efficiency of pinpointing plants' medicinal uses by more than 400 percent.

To mask the injustice and immorality of bioprospecting, agreements are made to compensate Third World countries for their contributions. For example, in 1992 Eli Lilly paid Shaman Pharmaceuticals, a major bioprospecting company, $4 million for exclusive worldwide marketing rights to anti-fungal drugs drawn from the knowledge of native healers. The Healing Forest Conservancy, Shaman's non-profit arm, will return a portion of its receipts to people and governments in the countries where Shaman works, although how much is never disclosed.

For Western corporations, indigenous systems of knowledge and indigenous rights do not exist. Thus, a publication of the pharmaceutical industry, which depends heavily on indigenous knowledge for many of its plant-based drugs, speaks of Third World biodiversity rights not as intellectual rights of people or as customary rights evolved over centuries, but as a newly asserted property right derived from a geographical accident. The most a developing country can claim for the drugs that are extracted by foreigners from their plants and animals is a geographical fee.[9] Yet some analysts propose that businesspeople, scientists, and lawyers meet to negotiate agreements. Neither the governments nor the people of countries rich in biodiversity figure in the thinking about bioprospecting contracts.[10]

One of the more publicized efforts was the 1991 agreement between Merck Pharmaceuticals and INBio, the National Biodiversity Institute of Costa Rica. Merck agreed to pay $1 million for the right to keep and analyze plant samples gathered from national Costa Rican rain forest parks by INBio employees. These unconditional rights for prospecting by a multinational corporation with $4 billion a year in revenues in exchange for $1 million paid to a small conservation organization do not respect the rights of local communities or the government of Costa Rica. Moreover, the agreement is not with the people living in or near the national parks; they had no say in the deal, nor were they guaranteed any benefits. Nor is it with the national government. The agreement is between a transnational corporation and a conservation group, developed at the initiative of a leading U.S. conservation biologist, Dan Janzen.

The intention of the Merck–INBio agreement is to stop the free flow of resources from the South to the North. As Janzen states, the days of exploration and exploitation without payment of royalties to the host country are over. For Janzen, Costa Rica is a corporation with 50,000 sq. km. of land, on which there are 12,000 sq. km. of 'greenhouses' filled with 500,000 species. This corporation has 3,000,000 stockholders. At present, there is $1,500 worth of gross national product (GNP) per stockholder. Costa Ricans aspire to a standard of living that is normally associated with about $10,000–$15,000 worth of GNP.

With this worldview, INBio views commercial prospecting by multinationals as the solution. Yet those selling prospecting rights never had the rights to biodiversity in the first place, and those whose rights are being sold and alienated through the transaction have not been consulted or given a chance to participate.

Further, while prospecting fees could be used to build

scientific capacity in the Third World, what is actually being built is a facility for the corporation. The current value of the world market for medicinal plants derived from leads given by indigenous and local communities is estimated to be $43 billion. Of this, on certain occasions, a tiny fraction is paid as prospecting fees. Such payments are supposed to build research capacity in the source country. But when Merck supplied chemical extraction equipment to the University of Costa Rica, for example, Merck ensured that it would have exclusive commercial use of the facility. This capacity-building is thus held captive by the financing corporation, and is not available for the wider national interest in the source country.

Another problem with biodiversity prospecting is that collections are often made as part of a scientific exchange in which the scientific bodies involved have links with corporations. Since scientific exchange takes place freely in the public domain, but the commercial interests exploiting the collections and screening have proprietary interests in developing products protected by IPRs, a major asymmetry of rights exists in biodiversity prospecting arrangements.

In other cases, indigenous communities are being asked to patent their knowledge in collaboration with Western corporations. The capital, however, comes from Western institutions and the rights are immediately transferred to powerful commercial interests, who control capital and the market. Drawing a few isolated groups or individuals into the gold rush for patents on life forms is becoming essential because the social movements saying 'no' to patents in the biodiversity domain are growing.

Does the patenting route protect indigenous knowledge? Protection of indigenous knowledge implies the continued availability and access to it by future generations in their everyday practices of health care and agriculture. If the economic organization that emerges on the basis of patents displaces the indigenous lifestyles and economic systems, indigenous knowledge is not being protected as a living heritage. If we recognize that the dominant economic system is at the root of the ecological crisis because it has failed to address the ecological value of natural resources, expanding the same economic system will not protect indigenous knowledge or biodiversity.

We need a transition to an alternative economic paradigm that does not reduce all value to market prices and all human activity to commerce.

Ecologically, this approach involves the recognition of the value of diversity in itself. All life forms have an inherent right to life; that should be the overriding reason for preventing species' extinction.

At the social level, the values of biodiversity in different cultural contexts need to be recognized. Sacred groves, sacred seeds, and sacred species have been the cultural means for treating biodiversity as inviolable, and present us with the best examples of conservation. Community rights to biodiversity, and farmers' and indigenous peoples' contributions to the evolution and protection of biodiversity, also need to be recognized by treating their knowledge systems as futuristic, not primitive. In addition, we need to recognize that non-market values, such as providing meaning and sustenance, should not be treated as secondary to market values.

At the economic level, if biodiversity conservation is to be aimed at conserving life rather than profits, the incentives given to biodiversity destruction and the penalties that have become associated with biodiversity conservation have to be removed. If a biodiversity framework guides economic thinking rather than the other way around, it becomes evident that the so-called high production of homogenous and uniform systems is an artificial measure, maintained through public subsidies. Productivity and efficiency need to be redefined to reflect the multiple input and multiple output systems that characterize biodiversity.

In addition, the perverse logic of financing biodiversity conservation by a small percentage of profits generated by biodiversity destruction amounts to licensing destruction, and reduces conservation to an exhibit, not a basis of living and producing.

Neither ecological sustainability nor livelihood sustainability can be ensured without a just resolution of the issue of who controls biodiversity. Until recent times, local communities, especially women, have used, developed, and conserved biological diversity, and have been the custodians of the biological wealth of this planet. Their control, their knowledge, and their rights need to be strengthened if the foundations of biodiversity conservation are to be strong and deep. This strengthening has to be done through local as well as national and global action.

The globalization of patent and IPR regimes is an expansion of the economic paradigm that has caused ecological destruction and contributed to the disappearance of species. When indigenous communities are brought into this paradigm, there is an irreversible destruction of the cultural diversity that could have provided values for another form of economic organization.

Taking knowledge from indigenous communities through bioprospecting is only the first step toward developing an IPR-protected industrial system that must eventually market commodities that use local knowledge as an input, but are not based on the ethical, epistemological, or ecological organization of that knowledge system. Producers of such commodities use biodiversity fragments as raw material to produce biological products protected by patents that displace biodiversity and indigenous knowledge, both of which they have exploited.

The issues of equity, fairness, and compensation need to be assessed in a systemic way, both at the level of taking indigenous knowledge and at the level of later pushing it out through aggressive marketing of industrialized products in medicine and agriculture. Key questions need to be asked. Is it right to displace the sources of alternative production and organization? Can such destruction be fully compensated? Can the planet, and the diverse communities that inhabit it, afford to have biodiversity and alternative lifestyles swallowed up as raw material for a centralized, global corporate culture that can only produce cultural and biological uniformity?

Patents, in the ultimate analysis, are systems of protection for capital investment, without the ability to control capital. As such, they protect neither people nor knowledge systems.

Bioprospecting does not respect the rights of people and communities who do not want the commons enclosed. Yet there are alternatives to bioprospecting.

Recovery of the Biodiversity Commons

There is a growing popular ecological movement to defend agricultural and medicinal biodiversity as well as peoples' knowledge. The protection and recovery of the biodiversity commons is, first and foremost, a political and social movement that recognizes the creativity intrinsic to the diversity of life forms. It calls for common property regimes in the ownership and utilization of biodiversity. Further, it works toward an intellectual commons—a public domain in which knowledge of biodiversity's utility is not commodified.

The first public demonstration of the positive assertion of the recovery of the biodiversity commons took place in India on Independence Day, August 15, 1993, when farmers declared that their knowledge is protected by *Samuhik Gyan Sanad* (collective intellectual rights). According to the farmers, any company using local knowledge or local resources without the permission of local communities is engaging in intellectual piracy, as in the case of the patents on neem.

The concepts were further developed in 1993 by an interdisciplinary team of experts from the Third World Network, an international group of Third World individuals and organizations. The positive assertion of collective intellectual property rights (CIRs) creates an opportunity to define a *sui generis* system of rights centred on the role of farmers in protecting and improving plant genetic resources.

Their effectiveness needs to be reinterpreted to account for the specific context of different countries. Only then would the diversity of IPR systems become a possibility; legal diversity, in turn, protects the biological and cultural diversity of peasant societies across the Third World. IPR diversity that has room for a plurality of systems, including regimes based on CIRs, would reflect different styles of knowledge generation and dissemination in different contexts. Alongside a positive protection system for farmers' rights as plant breeders, *sui generis* systems could evolve common rights in the domain of indigenous medical systems.

Additionally, the relationship between CIR systems that reflect Third World peoples' concerns and knowledge, and IPR regimes that have evolved with the Western bias toward individualized and juridical application procedures unsympathetic to rural societies, needs to be developed. A *sui generis* system must effectively prevent the systematic exploitation of Third World biological resources and knowledge, while it maintains the free exchange of knowledge and resources among Third World farming communities.

Sui generis systems that protect CIRs must necessarily be based on 'biodemocracy'—the belief that all knowledge and production systems using biological organisms have equal validity. In contrast, the TRIPs agreement is based on the concept of 'bioimperialism'—the belief that only the knowledge and production of Western corporations need protection. If unchallenged, TRIPs will become an instrument for displacing and dispensing with the knowledge, resources, and rights of Third World peoples, especially those who depend on biodiversity for their livelihoods, and who are the original owners and innovators in the utilization of biodiversity.

Legalizing Biopiracy

The TRIPs agreement of GATT is not the result of democratic negotiations between the larger public and commercial interests or between industrialized countries and the Third World. It is the imposition of values and interests by Western transnational corporations on the diverse societies and cultures of the world.

The framework for the TRIPs agreement was conceived and shaped by three organizations—the Intellectual Property Committee (IPC), Keidanren, and the Union of Industrial and Employees Confederations (UNICE). IPC is a coalition of 12 major U.S. corporations: Bristol Myers, DuPont, General Electric, General Motors, Hewlett Packard, IBM, Johnson & Johnson, Merck, Monsanto, Pfizer, Rockwell, and Warner. Keidanren is a federation of economic organizations in Japan, and UNICE is recognized as the official spokesperson for European business and industry.

The transnational corporations have a vested interest in the TRIPs agreement. For example, Pfizer, Bristol Meyers, and Merck already have patents on Third World biomaterials collected without payment of royalties.

Together, these groups worked closely to introduce intellectual property protection into GATT.

James Enyart of Monsanto, commenting on the IPC strategy, states:

> Since no existing trade group or association really filled the bill, we had to create one . . . Once created, the first task of the IPC was to repeat the missionary work we did in the U.S. in the early days, this time with the industrial associations of Europe and Japan to convince them that a code was possible . . . We consulted many interest groups during the whole process. It was not an easy task but our Trilateral Group was able to distill from the laws of the more advanced countries the fundamental principles for protecting all forms of intellectual property . . . Besides selling our concepts at home, we went to Geneva where [we] presented

[our] document to the staff of the GATT Secretariat. We also took the opportunity to present it to the Geneva based representatives of a large number of countries . . . What I have described to you is absolutely unprecedented in GATT. Industry has identified a major problem for international trade. It crafted a solution, reduced it to a concrete proposal and sold it to our own and other governments . . . The industries and traders of world commerce have played simultaneously the role of patients, the diagnosticians and the prescribing physicians.[11]

By usurping all these roles from diverse social groups, commercial interests have displaced ethical, ecological, and social concerns from the substance of the TRIPs agreement. Prior to the Uruguay Round of GATT, which concluded in 1993, IPRs were not covered. Each country had its own national IPR laws to suit its ethical and socioeconomic conditions. The major thrust for internationalizing IPR laws was given by the transnational corporations (TNCs). Even though IPRs are only a statutory right, the TNCs have naturalized them. They have then used GATT to protect what they define as their 'rights' as owners of intellectual property. As stated in the 1988 industry paper 'Basic Framework for GATT Provisions on Intellectual Property', co-authored by IPC, Keidanren, and UNICE:

Because national intellectual property protection systems differ, intellectual property right owners spend a disproportionate amount of time and resources to acquire and defend rights. These owners also find that the exercise of intellectual property rights is encumbered by laws and regulations limiting market access or the ability to repatriate profits.[12]

All of the undesirable elements in the Patent (Amendment) Bill are to be found in this 1988 industry paper. These include expanding the life, subject matter, and scope of product patents, while shrinking the requirements for the working of a patent and compulsory licensing. While the Indian Patent

Act of 1970 does not allow product patents in pharmaceuticals and agrochemicals, the Patent (Amendment) Bill, introduced in 1995 by the Indian government to implement GATT TRIPs, but ultimately rejected, allows product patent applications and granting of exclusive marketing rights. This push for product patents is clearly expressed in the 'Basic Framework' paper:

> Some countries which grant protection for mechanical and electrical services deny protection for new substances. In the case of chemicals, pharmaceuticals and agrochemicals, for example, some countries permit only a patent for a specific process of making the product, while others provide protection for the product only when made by the process [product-by-process protection]. Chemical substances, however, can almost always be made in a variety of ways, and it is rarely feasible to patent all such routes. Where the invention resides in a new valuable chemical substance, a process patent is therefore simply an invitation to imitators to manufacture the chemical substance by another route, usually a relatively straightforward exercise for a competent chemist.[13]

Similarly, the Indian Patent Act has strong compulsory licensing clauses to ensure that the fundamental right of the public to food and medicine is not ignored due to the profit motive. TNCs, however, view this protection of the public interest as discrimination. As they state:

> Grant of an exclusive right is an essential element of an effective patent system. However, some countries subject patents in a particular field to compulsory licensing to third parties on demand. Food, medicines and sometimes agrochemicals are particular targets for this form of discrimination. This results in undue injury to the rights of its owner.[14]

In the TNC approach, the injury caused by exclusive marketing rights and monopolies to the fundamental human rights of citizens to be able to meet their basic needs is of no consequence. The TNCs define all public interest elements in IPR regimes, such as systems of working requirement and compulsory licenses, as an abuse. According to them, commercial reality should be the only consideration. Ethical limits as well as social and economic imperatives are mere barriers to their commercial expansion.

Under the one-sided influence of TNCs, life forms have been included under the subject matter of patentability. Since most of the companies in IPC have interests in chemicals, pharmaceuticals, agrochemicals, and the new biotechnologies, they demanded the inclusion of biological organisms under patent protection. As stated in the 'Basic Framework':

> Biotechnology, the use of micro-organisms to make products, represents a related field in which patent protection has not kept pace with the rapid strides being made in health, agriculture, waste disposal and industry. The products of biotechnology include the building blocks to make genes, hybridomas, monoclonal antibodies, enzymes, chemicals, micro-organisms and plants. Although biotechnology has attracted widespread attention, many countries withhold the effective patent protection needed to justify investment in relevant research and development. Such protection should be afforded for both biotechnology processes and products, including microorganisms, parts of micro-organisms (plasmids and other vectors) and plants.[15]

The issue of the patentability of life is not merely a trade-related issue: it is primarily an ethical and ecological issue intimately related to the social injustice of biopiracy. If implemented, the TRIPs agreement could have tremendous implications for the health of the environment as well as for the conservation of biodiversity.

Chapter Five

Tripping Over Life

Diversity is the key to sustainability. It is the basis of mutuality and reciprocity—the 'law of return' based on the recognition of the right of all species to happiness and freedom from suffering. Yet the law of return based on freedom and diversity is being replaced by the logic of return on investments. Genetic engineering, even while preying on the world's biological diversity, threatens to aggravate the ecological crisis through the expansion of monocultures and monopolies.

The TRIPs agreement of GATT, by allowing for monopolistic control of life forms, has serious ramifications for biodiversity conservation and the environment. Article 27.5.3(b) of the TRIPs agreement states:

> Parties may exclude from patentability plants and animals other than micro-organisms, and essentially biological processes for the production of plants or animals other than non-biological and microbiological processes. However, parties shall provide for the protection of plant varieties either by patents or by an effective *sui generis* system or by any combination thereof. This provision shall be reviewed four years after the entry into force of the Agreement establishing the WTO.

The most significant ecological impacts of TRIPs are related to changes in the ecology of species interactions that will be brought about as a result of commercial releases of patented and genetically engineered organisms (GEOs). TRIPs also affect biodiversity rights, which, in turn, lead to changes in the sociocultural context of conservation. Some of these impacts are:

1. The spread of monocultures, as corporations with IPRs attempt to maximize returns on investments by increasing market shares.

2. An increase in chemical pollution, as biotechnology patents create an impetus for genetically engineered crops resistant to herbicides and pesticides.

3. New risks of biological pollution, as patented genetically engineered organisms are released into the environment.

4. An undermining of the ethics of conservation, as the intrinsic value of species is replaced by an instrumental value associated with intellectual property rights.

5. The undermining of traditional rights of local communities to biodiversity, and hence a weakening of their capacity to conserve biodiversity.

The Spread of Monocultures

The conservation of biodiversity requires the existence of diverse communities with diverse agricultural and medical systems that utilize diverse species *in situ*. Economic decentralization and diversification are necessary conditions for biodiversity conservation.

The globalized economic system dominated by TNCs, in which TRIPs are embedded and consolidated even further, creates the conditions for the spread of uniformity and the destruction of diversity.

Diverse crop varieties have evolved according to different environmental conditions and cultural needs. The genetic variability of these varieties is insurance against pests, disease, and environmental stress. This resilience is enhanced by traditional agricultural practices, like mixed cropping.

Corporations that obtain IPRs for plants or animals need

to maximize their return on investment which, in turn, creates pressure to maximize their market share. The same variety of crop or livestock is therefore spread worldwide, leading to the displacement of hundreds of local varieties of crops and breeds of livestock. The spread of monocultures and the destruction of diversity is an essential aspect of global markets protected by IPRs.

Monocultures, however, are ecologically unstable, inviting disease and pests. For instance, in 1970-71, the United States experienced a maize blight epidemic, which laid waste to 15 percent of the nation's crop, because of genetic uniformity. Eighty percent of the hybrid maize in the United States in 1970 was derived from a single, sterile male line and contained T. cytoplasm, which made the plants vulnerable to the maize blight fungus *Helminthosporum maydis*. It left ravaged maizefields with withered plants, broken stalks, and malformed or completely rotten cobs with a greyish powder. Plant breeders and seed companies used T. cytoplasm only because it fostered quick and profitable production of high-yielding, hybrid corn seed. As a University of Iowa pathologist wrote after the blight: "Such an extensive, homogenous acreage is like a tender prairie waiting for a spark to ignite." [1]

According to a 1972 National Academy of Sciences study on the genetic vulnerability of major crops:

> The corn crop fell victim to the epidemic because of a quirk in the technology that had designed the corn plants of America, until, in one sense, they had become as alike as identical twins. Whatever made one plant susceptible made them all susceptible. [2]

The spread of monocultures of high-yielding varieties in agriculture and fast-growing species in forestry has been justified on grounds of increased productivity. The technological transformation of biodiversity—as well as the granting of IPRs and patent monopolies—is justified in the language of improvement and increase of economic value. These terms are not neutral, however; they are contextual and value-laden. Improving tree species means one thing for a paper corporation, which needs pulping wood, and another for a peasant, who needs fodder and green manure. Improving crop species means one thing for a processing industry and something totally different for a self-provisioning farmer. Thus, Cargill—the largest grain trader and the fourth largest seed corporation—has demanded IPRs to protect its investment, claiming this is a social necessity because it is supposed to benefit farmers.

Yet farmers in Karnataka have had the opposite experience. In 1992, when Cargill first entered the Indian seed market, its sunflower seeds were a total failure. Instead of the promised 1,500 kilograms per acre, they yielded only 500 kilograms per acre.

Similarly, Cargill's hybrid sorghum has led to a decline in farmers' income due to the much higher cost of purchased inputs. In 1993 in Karnataka, India, farmers' cost of production with Cargill's hybrid sorghum was Rs. 3,230 per acre, and their income was Rs. 3,600 per acre, according to a survey by the Research Foundation for Science, Technology, and Natural Resource Policy. In contrast, according to the survey, farmers' cost of production with indigenous seeds was Rs. 300 per acre, and their income was Rs. 3,200 per acre. The hybrid seeds returned just Rs. 370 per acre, while native seeds returned Rs. 2,900 per acre.

Intensification of Chemical Pollution

Patent protection as guaranteed under TRIPs will encourage biotechnological interventions and accelerate the release of genetically engineered organisms. While the sales appeal of genetic engineering is through the 'green' image of chemical-free agriculture, most agricultural applications of biotechnology focus on increased use of agrochemicals. The impact of these applications will be higher in the Third World not only because the native biodiversity is higher, but because livelihoods are more dependent on this diversity.

Most research and innovation in agricultural biotechnology is being undertaken by chemical multinationals, such as Ciba-Geigy, ICI, Monsanto, and Hoechst. Their immediate strategy is to increase the use of pesticides and herbicides by developing pesticide- and herbicide-tolerant crop varieties. Twenty-seven corporations are working on virtually all the major food crops to develop herbicide resistance. For the seed-chemical multinationals, this makes commercial sense, since it is cheaper to adapt the plant to the chemical than the chemical to the plant. The cost of developing a new crop variety rarely tops $2 million, whereas the cost of a new herbicide exceeds $40 million.

Herbicide and pesticide resistance will also increase the integration of seeds and chemicals, and hence the control of agriculture by multinational corporations. A number of major agrochemical companies are developing plants with resistance to their own brand of herbicide. Soya beans have been made resistant to Ciba-Geigy's Atrazine herbicide, thus increasing annual sales by $120 million. Research is being done to develop crop plants that are resistant to other herbicides, such as DuPont's Gist and Glean and Monsanto's Round-Up, which are lethal to most herbaceous plants and therefore cannot be applied directly to crops. The successful development and sale of crop plants resistant to brand-name herbicides will result in the further economic concentration of the agro-industry market, increasing the power of transnational companies. The Ministry of Environment in Denmark, in its environmental risk assessment of herbicide-resistant agricultural crops, stated:

> The present case is concerned with a plant, which exists as a weed in other crops and which is closely related to wild species. As described below, there may be an exchange of genes between oilseed rape and related species. The spreading of resistance, especially combinations of resistance, will make it more difficult

to eradicate oilseed rape with minimal use of herbicides and the rape itself will also appear as a weed, which is difficult to control in other crops. Patterns of herbicide use probably change. In this particular case, resistance has also been introduced to a herbicide (Basta), which is characterized by being effective against practically all weed species of importance. It is, therefore, to be expected that the transfer of resistant genes to weeds will cause a gradual spreading of resistance to this agent and is thus likely to result in increased and wider use of herbicides.

New Forms of Biological Pollution

Strategies to genetically engineer herbicide resistance, which are destroying useful species of plants, can end up creating superweeds. There is an intimate relationship between weeds and crops, especially in the tropics, where weedy and cultivated varieties have genetically interacted for centuries and hybridize freely to produce new varieties. Genes for herbicide tolerance, pest resistance, and stress tolerance, introduced into crop plants by genetic engineers, may be transferred to neighbouring weeds as a result of natural hybridization.[3] This, in turn, will lead to increased chemical use, with all the associated environmental risks.

The hazards of gene transfer to wild relatives are higher in the Third World, as these regions are home to most of the world's biodiversity. As the U.S. Academy of Sciences guide, *Field Testing Genetically Modified Organisms*, states:

Temperate North America, especially the United States, includes the home ranges for very few crops, as U.S. agriculture is based largely on crops of foreign origin. This paucity of crops derived from North American sources means there will be relatively few opportunities for hybridization between crops and wild relatives in the United States. The incidence of hybridization between genetically modified crops and wild relatives can be expected to be lower here than in Asia Minor, South East

Asia, the Indian subcontinent, and South America, and greater care may be needed in the introduction of genetically modified crops in those regions.[4]

Genetically engineered organisms also create new risks of biological pollution. As Dr. Peter Wills has stated: "There are serious, but unpredictable consequences, of converting the phytogenetic tree of DNA into an interspecies network."

Recent experiments have established that the large-scale transfer of engineered traits into related species are a real possibility.

Biological pollution can also occur when non-genetically engineered species are introduced into ecosystems. For instance, in 1970, the Blue Tilapia was introduced in Lake Effie in Florida. It constituted less than 1 percent of the total weight (biomass) of fish in the lake. By 1974, the Blue Tilapia had dominated other species and accounted for 90 percent of the total biomass.

In another case, in the late 1950s the British introduced Nile perch into Lake Victoria in East Africa to increase fish production. The indigenous species were small and diverse, including some 400 species of haplochromines, which weigh about one pound each but made up 80 percent of the fish biomass in the lake. The Nile perch is a carnivorous fish that can grow to six feet in length and weigh 150 pounds.

During the next 20 years, nothing happened. In the early 1980s, however, Nile perch took over Lake Victoria. Before 1980, it had comprised about 1 percent of the catch; by 1985, it made up 60 percent. Fish biomass in the lake shifted from 80 percent haplochromines to 80 percent Nile perch. Haplochromines now account for less than 1 percent of the fish biomass. Scientists estimate that half of the original 400 species of haplochromines in Lake Victoria are extinct.

Recently, there has been a decline in the Nile perch catch. Those that are caught are similar and many have juvenile Nile perch in their stomachs. When species start to feed on their own offspring, it is a sign of ecological instability and a result-ing break in the food chain.

One final example is the introduction of opossum shrimp to improve production of kakonee salmon in Manitoba's Flathead Lake. It had the reverse effect, actually leading to a decline in kakonee salmon. The opossum shrimp turned out to be a voracious predator of zooplankton, an important source of food for the salmon. After the introduction of opos-sum shrimp, the zooplankton population was reduced to 10 percent of its former level. The spawning of salmon declined from 118,000 to 26,000 in 1986, to 330 in 1987, to 50 in 1989. The catch declined from above 100,000 in 1985, to 600 in 1987, and zero in 1988 and 1989.

Genetically engineered organisms that establish self-sustain-ing populations in nature will need to be assessed for their impact on other organisms. For this, a reductionist molecular biology is inadequate. It can classify the genetic composition of a species, but the ecological impact is determined by the nature and magnitude of the interaction between genes, their expression in different organisms, and the environment. Ecological questions need to be raised about the natural interactions of the host species with other organisms, its role in ecosystem processes, and the related consequences of pos-sible differences exhibited by transgenic organisms. Transgenic fish released into the environment may be resistant to popu-lation regulating factors, such as disease, parasitism, and preda-tion. They may also pass their transgenes to related species, and change the nature of predator-prey relationships.[5]

Even though GEOs exhibit little environmental impact in the short term, this is no reason for complacency about issues of biosafety. Indeed, many GEOS may never threaten ecosys-tems. Yet those few that do pose serious threats of biological pollution, especially over the long term.

Undermining the Ethics of Conservation

Intellectual property rights over life forms are an extreme expression of an instrumental view of other species, which conservation ethics views conversely as having intrinsic worth. The intrinsic value of other species presents humans with a prima-facie duty and responsibility not to use organisms as lifeless, valueless, structureless objects. When the intrinsic value of species is replaced by the instrumental value built into IPR claims, the ethical basis for biodiversity conservation and compassion for other species is undermined.

This compassion is the basis of ancient religions such as Buddhism, Jainism, and Hinduism, as well as new movements, such as the protests against live calf exports and hunting in the United Kingdom. Ancient religions and new movements both reinforce the belief in the intrinsic value of species.

Article 2 of TRIPs allows for the exclusion of patents on life on ethical and ecological grounds. Most groups concerned with these ethical issues, however, do not even know that trade treaties could have implications for their fundamental ethical principles. It should, therefore, be obligatory that before TRIPs implementation takes place, the implications for life forms are publicized and the views of diverse groups are heard.

Ron James, a spokesperson for the biotechnology industry and the 'maker' of Tracy, is vociferous in asserting that patents are not a moral issue because they do not confer a right to do something. They are ethically neutral; they merely exclude others from using an innovation. This ethical evasion, however, does not address the fact the IPRs are claims to intellectual property, and patents give exclusive rights to patent holders to make products based on these claims. In essence, patents are ownership claims on the basis of making something novel.

Certainly, the idea of owning life is not new; people own their pets and farmers own their livestock. Yet IPRs create a new concept of ownership. It is not just the implanted gene, or one generation of animals, that is being claimed as intellectual property, but the reproduction of the entire organism, including future generations covered by the life of the patent.

Alienation of Local Rights

Biodiversity conservation depends on the rights of local communities to enjoy the fruits of their efforts. Alienation of these rights rapidly leads to the erosion of biodiversity, which in turn threatens ecological survival and economic wellbeing. IPRs in the areas of biodiversity and life forms are not merely a creation of new rights; they also involve a re-writing of the traditional rights that enabled local communities to be the keepers of biodiversity, with a stake in its replenishment and utilization. IPRs in seeds, plant material, and indigenous knowledge systems alienate the rights of local communities and undermine the stake they have in the protection of biodiversity.

For example, when village forests were enclosed by the British in colonial India, the local people were denied their traditional rights to forest resources. While colonial forest policy became a license for massive deforestation, the local people were often blamed for the devastation. As G. B. Pant observes:

> The tale about the denudation of forests by the hillman was repeated *ad nauseam* in season and out of season by those in power so much so that it came to be regarded as an article of faith ... By way of vindication of the forest policy, it is claimed by its advocates that in the pre-British days the people had neither any rights in the soil nor in the forests.
>
> The policy of the Forest Department can be summed up in two words, namely, encroachment and exploitation. The Government has gone on pushing forward, extending its own sphere and scope and simultaneously narrowing down the orbit

of the rights of the people . . . The memory of the San assi boundaries [1880 pre-demarcation] is green and fresh in the mind of every villager and he cherishes it with a feeling bordering on reverence; he is simply unable to see his way to accepting the claim of the Government to the benap [unmeasured under settlement records] lands comprised within his village boundaries and regards every advance in that line as nothing short of encroachment and intrusion. Let the San assi boundaries be vested with their real character instead of being looked upon as merely nominal, and, to remove misgivings, let the areas enclosed within these boundaries be declared as the property of the villagers and all the benap lands included within these areas be restored to the village community, subject to such conditions as to impartibility, etc., as may be desirable in the public interest. It is a matter of common knowledge that a large number of memorials were sent by the villagers at their own instance, about the year 1906, asking the Government to restore the areas within the San assi boundaries to them: the unsophisticated villager spontaneously reiterates the same demands today. This is the minimum demand of the people and there seems to be no other rational and final solution. The simple fact should not be forgotten that man is more precious in this earth than everything else, the forests not excepted, and, also, that coercion is no substitute for reason, and, however stringent and rigid the laws may be, the forests cannot be preserved in the midst of seething discontent against the unanimous wishes and sentiments of the people . . . The collective intelligence of a people cannot be treated with contempt, and even if it be erratic, it can come round only by being allowed an opportunity of realizing its mistake. If the village areas are restored to the villagers, the causes of conflict in antagonism between the forest policy and the villagers will take the place of the present distrust, and the villager will begin to protect the forests even if such protection involves some sacrifice or physical discomfort.

This alienation of local rights was the basis of the Forest *Satyagraha* of the 1930s, which erupted across the country and in the Himalayas, in Central India, and in the western Ghats. M. K. Gandhi developed *satyagraha* ('struggle for the truth') as a form of peaceful non-cooperation with unjust laws and regimes. As G. S. Halappa reports about the Jungle *Satyagraha* in the western Ghats:

> The government began to arrest the *satyagrahis*, who had come from outside, and a few important local leaders. The latter awakened the women to action . . . The jungle *satyagraha* could not be put down by force for the people of whole villages would move out in thousands and would vie with one another in getting arrested.[7]

When seeds are covered by patents or plant breeder rights, and market forces combine with IPR protection to shift seed supplies from the farmer to the corporation, farmers' rights as breeders and innovators are undermined and the incentives for on-farm conservation are undone, leading to rapid genetic erosion.

In 1992, on Gandhi's birthday, the Seed *Satyagraha* was launched in India to resist the alienation through the TRIPs treaty of farmers' rights to seed and agricultural biodiversity. The alienation of local rights has also been identified as the primary reason for biodiversity erosion in Ethiopia, according to the National Conservation Strategy:

> Perhaps the most important policy and regulatory interventions in terms of their negative impacts on the environment were those impositions, which increasingly and cumulatively eroded the rights of individuals and communities to use and manage their own resources . . . Because farmers and communities did not have any control over trees which they might plant, either they did not plant any at all, or when coerced to plant did not maintain or care for them. In this way many community wood-lots planted with great physical effort resulted in little gain.[8]

Agricultural biodiversity has been conserved only when farmers have total control over their seeds. Monopoly rights regimes for seeds, either in the form of breeders' rights or patents, will have the same impact on *in situ* conservation of plant genetic resources as the alienation of rights of local communities has had on the erosion of tree cover and grasslands in Ethiopia, India, and other biodiversity-rich regions.

Chapter Six

Making Peace with Diversity

In this time of 'ethnic cleansing', as monocultures spread throughout society and nature, making peace with diversity is fast becoming a survival imperative.

Monocultures are an essential component of globalization, which is premised on homogenization and the destruction of diversity. Global control of raw materials and markets makes monocultures necessary.

This war against diversity is not entirely new. Diversity has been threatened whenever it has been seen as an obstacle. Violence and war are rooted in treating diversity as a threat, a perversion, a source of disorder. Globalization transforms diversity into a disease and deficiency because it cannot be brought under centralized control.

Homogenization and monocultures introduce violence at many levels. Monocultures are always associated with political violence—the use of coercion, control, and centralization. Without centralized control and coercive force, this world filled with the richness of diversity cannot be transformed into homogeneous structures, and the monocultures cannot be maintained. Self-organized and decentralized communities and ecosystems give rise to diversity. Globalization gives rise to coercively controlled monocultures.

Monocultures are also associated with ecological violence—a declaration of war against nature's diverse species. This violence not only pushes species toward extinction, but controls and maintains monocultures themselves. Monocultures are non-sustainable and vulnerable to ecological breakdown.

Uniformity implies that a disturbance to one part of a system is translated into a disturbance to other parts. Instead of being contained, ecological destabilization tends to be amplified. Sustainability is ecologically linked to diversity, which offers the self-regulation and multiplicity of interactions that can heal ecological disturbance to any part of a system.

The vulnerability of monocultures is well illustrated in agriculture. For example, the Green Revolution replaced thousands of local rice varieties with the uniform varieties of the International Rice Research Institute (IRRI). JR-8, released in 1966, was hit in 1968-69 by bacterial blight and attacked by timgro virus in 1970-71. In 1977, IR-36 was bred for resistance to eight major diseases, including bacterial blight and timgro. But being a monoculture it was vulnerable to attack by two new viruses, 'ragged stunt' and 'wilted stunt'. [1]

The miracle varieties displaced the diversity of traditionally grown crops, and through the erosion of diversity the new seeds became a mechanism for introducing and fostering pests. Indigenous varieties are resistant to local pests and diseases. Even if certain diseases occur, some of the strains may be susceptible, but others will have the resistance to survive.

What happens in nature also happens in society. When homogenization is imposed on diverse social systems through global integration, region after region starts to disintegrate. The violence inherent to centralized global integration, in turn, breeds violence among its victims. As conditions of everyday life become increasingly controlled by outside forces and systems of local governance decay, people cling to their diverse identities as a source of security in a period of insecurity. Tragically, when the source of their insecurity is so remote that it cannot be identified, diverse peoples who have lived peacefully together start to look at each other with fear. Markings of diversity become cracks of fragmentation; diversity then becomes the justification for violence and war, as we have seen in Lebanon, India, Sri Lanka, Yugoslavia, Sudan, Los Angeles, Germany, Italy, and France. As local and national systems of governance break down due to the pressures of globalization, local elites attempt to cling to power by manipulating the ethnic or religious feelings that emerge as a backlash.

In a world characterized by diversity, globalization can only be realized by ripping apart society's plural fabric along with its capacity to self-organize. At the political and cultural level, it is this freedom to self-organize that Gandhi saw as the basis of interaction between different societies and cultures. "I want the cultures of all lands to be blown about as freely as possible, but I refuse to be blown off my feet by any," he said.

Globalization is not the cross-cultural interaction of diverse societies; it is the imposition of a particular culture on all of the others. Nor is globalization the search for ecological balance on a planetary scale. It is the predation of one class, one race and often one gender of a single species on all of the others. The 'global' in the dominant discourse is the political space in which the dominant local seeks global control, freeing itself of responsibility for the limits arising from the imperatives of ecological sustainability and social justice. In this sense, the 'global' does not represent a universal human interest; it represents a particular local and parochial interest and culture that has been globalized through its reach and control, its irresponsibility and lack of reciprocity.

Globalization has occurred in three waves. The first wave was the colonization of America, Africa, Asia, and Australia by European powers over 1,500 years. The second imposed a Western idea of 'development' during the postcolonial era of the past five decades. The third wave of globalization, unleashed approximately five years ago, is known as the era of 'free trade'. For some commentators, this implies an end to history; for the Third World, it is a repeat of history through recolonization. The impact of each wave of globalization is cumulative, even as it creates discontinuity in the dominant metaphors and actors. And each time a global order has tried to wipe out diversity and impose homogeneity, disorder and disintegration have been induced, not removed.

Globalization 1: Colonialism

When Europe first colonized the diverse lands and cultures of the world, it also colonized nature. The transformation of the perception of nature during the industrial and scientific revolutions illustrates how 'nature' was transformed in the European mind from a self-organizing, living system to a mere raw material for human exploitation, needing management and control.

'Resource' originally implied life. Its root is the Latin *resurgere*, or 'to rise again'. In other words, resource means self-regeneration. The use of the term 'resource' for nature also implied a relationship of reciprocity between nature and humans.[2]

With the rise of industrialism and colonialism, a shift in meaning took place. 'Natural resources' became inputs for industrial commodity production and colonial trade. Nature was transformed into dead and manipulable matter. Its capacity to renew and grow had been denied.

The violence against nature, and the disruption of its delicate interconnections, was a necessary part of denying its self-organizing capacity. And this violence against nature, in turn, translated into violence in society.

Anything not fully managed or controlled by European men was seen as a threat. This included nature, non-Western societies, and women. What was self-organized was considered wild, out of control, and uncivilized. When self-organization is perceived as chaos, it creates a context to impose a coercive and violent order for the betterment and improvement of the 'other', whose intrinsic order is then disrupted and destroyed.

Most non-Western cultures have regarded the wild as sacred, viewing its diversity as a source of inspiration for democracy and freedom. Rabindranath Tagore, India's national poet, writing in Tapovan at the peak of the independence movement, saw democracy in society as derived from the principles of diversity in nature, whose highest expression is found in the forest. The diverse processes of renewal that are always at play in the forest—varying from species to species, from season to season, in sight, sound, and smell—have fuelled the culture of Indian society. The unifying principle of life in diversity, of democratic pluralism, thus became the principle of Indian civilization.[3]

Whenever Europeans 'discovered' the native peoples of America, Africa, or Asia, they identified them as savages in need of redemption by a superior race. Even slavery was justified on these grounds. To carry Africans into slavery was seen as an act of benevolence, transporting them from the 'endless night of savage barbarism' into the embrace of a 'superior civilization'.

The West's fear of the wild and its associated diversity is closely linked to the imperative of human domination, and the control and mastery of the natural world. Thus Robert Boyle, the famous scientist who was also Governor of the New England Company in the 1760s, saw the rise of mechanical philosophy as an instrument of power not just over nature, but also over the original inhabitants of America. He explicitly declared his intention of ridding the New England Indians of their absurd notions about the workings of nature. Boyle attacked their perception of nature "as a kind of goddess", and argued that "the veneration, wherewith men are imbued for what they call nature, has been a discouraging impediment to the empire of man over the inferior creatures of God."[4] The concept of the 'empire of man' was thus substituted for the 'earth family', where humans are included in the pluralism of nature's diversity.

This conceptual diminution was essential to the projects of colonization and capitalism. The concept of an earth family excluded the possibilities of exploitation and domination; a denial of the rights of nature as well as societies that revere nature was necessary in order to facilitate uncontrolled exploitation and profits.

Diversity, being a threat, had to be wiped out of a world-view where European men were the measure of being human and having human rights. As A. W. Crosby observes:

> Again and again, during the centuries of European imperialism, the Christian new that all men are brothers was to lead to persecution of non-Europeans—he who is my brother sins to the extent that he is unlike me.[5]

All brutality was sanctioned on the basis of the assumed superiority of European men and their exclusive status as fully human. As Basil Davidson observes, the moral justification for invading and expropriating the territory and possessions of other peoples was the assumed 'natural' superiority of Europeans to the 'tribes without law', the 'fluttered folk and wild'.[6]

Denying other cultures their rights on the basis of their difference from European culture was convenient for taking away their resources and wealth. The church authorized European monarchs to attack, conquer, and subdue non-believers, to capture their goods and their territories, and to transfer their lands and properties. Five hundred years ago, Columbus carried this worldview to the New World, and as a result of the first wave of globalization, millions of people and thousands of other living species lost their right to exist.

Globalization II: 'Development'

The war against diversity did not end with colonialism. The definition of entire nations of people as incomplete and defective Europeans was reincarnated in the 'development' ideology, which predicated their salvation on generous assistance and advice from the World Bank, the International Monetary Fund (IMF) and other financial institutions, and multinational corporations.

Development is a beautiful word, suggesting evolution from within. Until the middle of the 20th century, it was

synonymous with evolution as self–organization. But the ideology of development has implied the globalization of the
priorities, patterns, and prejudices of the West. Instead of
being self–generated, development is imposed. Instead of
coming from within, it is externally guided. Instead of contributing to the maintenance of diversity, development has
created homogeneity and uniformity.

The Green Revolution is a prime example of the development paradigm. It destroyed diverse agricultural systems
adapted to the diverse ecosystems of the planet, globalizing
the culture and economy of an industrial agriculture. It wiped
out thousands of crops and crop varieties, substituting them
with monocultures of rice, wheat, and maize across the Third
World. It replaced internal inputs with capital- and chemical-
intensive inputs, creating debt for farmers and death for
ecosystems.

The Green Revolution did not merely unleash violence
against nature. By creating an externally managed and glob-
ally controlled agriculture, it sowed the seeds for violence in
society.

Rural development in general, and the Green Revolution
in particular, assisted by foreign capital and planned by foreign
experts, were prescribed as means for peace by politically stabilizing rural areas and preventing areas outside of China from
falling under the influence of the Red Revolution. After two
decades, however, the invisible ecological, political, and cultural costs of the Green Revolution became apparent. At the
political level, the Green Revolution turned out to produce
rather than reduce conflict. At the material level, high yields
of commercial grain generated new scarcities at the ecosystem
level, in turn generating new sources of conflict. At the cultural level, the homogenization processes of the Green
Revolution led to the resurgence of ethnic and religious
identities.

The ecological and ethnic crises in the Third World can be
viewed as arising from a basic and unresolved conflict

between the demands of diversity, decentralization, and democracy on the one hand, and uniformity, centralization, and militarization on the other. Control over nature and people was an essential element of the centralized and centralizing strategy of the Green Revolution. Ecological breakdown in nature and the political breakdown of society were the implications of a policy based on tearing apart both nature and society.

The Green Revolution was based on the assumption that technology is a superior substitute for nature, and hence a means of producing growth unconstrained by nature's limits. Conceptually and empirically, viewing nature as a source of scarcity and technology as a source of abundance leads to technologies that create new scarcities in nature through ecological destruction. Green Revolution practices, for example, reduced the availability of fertile land and the genetic diversity of crops, thereby creating scarcity.

The Green Revolution's shift from cropping systems based on diversity and internal inputs to ones based on uniformity and external inputs did not merely change the ecological processes of agriculture. It also changed the structure of social and political relationships, from those based on mutual (though asymmetric) obligations—within the village—to relations between individual cultivators and their banks, seed and fertilizer agencies, food procurement agencies, and electricity and irrigation organizations. Atomized and fragmented, cultivators relating directly to the state and the market generated an erosion of cultural norms and practices. Further, since the externally supplied inputs were scarce, it set up conflict and competition between classes and between regions, and sowed the seeds of violence and conflict.

The centralized planning and allocation that made the Green Revolution possible affected not only people's lives, but it also affected their very idea of self. With government as referee, handing down decisions in all matters, each frustration became a political issue. In a context of diverse communities,

that centralized control led to communal and regional conflict. Every policy decision translated into the politics of 'we'
and 'they'—'we' have been unjustly treated, while 'they' have
gained privileges unfairly.

As Francine Frankel wrote in 1972 in *The Political
Challenge of the Green Revolution*:

> It is not too early, moreover, to consider one major implication
> of this analysis, namely that disruption is accelerated to so rapid
> a rate that the time available for autonomous re-equilibrating
> processes, even if such processes are operative ... is critically cur-
> tailed. Thus in the absence of countervailing initiatives, forces
> already in motion will push traditional societies in rural areas to
> a total breakdown.[8]

In 1972, the prediction of breakdown seemed far-fetched; yet
in 1984, two Sikh extremists assassinated Indira Gandhi. Two
thousand Sikhs were massacred in Delhi as a backlash. In
1986, 598 people were killed in Punjab; one year later, the
number was 1,544. In 1988, the number had risen to 3,000.

The rapid and large-scale introduction of Green
Revolution technologies dislocated social structures and
political processes at two levels. It created growing disparities
among classes, while also increasing the commercialization of
social relations. As Frankel observed, the Green Revolution
completely eroded social norms. "In those regions where the
new technology has been most extensively applied, it has
accomplished what a century of disruption under colonial
rule failed to achieve, the virtual elimination of the stabilizing
residuum of traditional society."

While Frankel had predicted social breakdown, she had
seen it as emerging from class conflict. Yet as the Green
Revolution unfolded, communal and ethnic aspects came to
the fore. Modernization and economic development may, as
in the case of Punjab, harden ethnic identities, provoking or
intensifying conflicts on the basis of religion, culture, or race.

To a large extent, the movements for regional, religious,

and ethnic revival are movements for the recovery of diversity in the context of homogenization. The paradox of separatism, however, is that it searches for diversity within a framework of uniformity. It is a search for identity in a structure based on erasure and erosion of identities. The shift from Sikh farmers demanding their rights to the demand for a separate Sikh state comes from the collapse of horizontally organized, diverse communities into atomized individuals linked vertically to state power through electoral politics.

The homogenization processes of development do not fully wipe away differences. Differences persist—not in the integrating context of plurality, but in the fragmenting context of homogenization. Positive pluralities give way to negative dualities, in competition with each other, contesting for the scarce resources that define economic and political power. Diversity is mutated into duality, into the experience of exclusion. The intolerance of diversity becomes a new social disease, leaving communities vulnerable to breakdown and violence, decay and destruction. The intolerance of diversity and the persistence of cultural differences sets one community against another in a context created by a homogenizing state, carrying out a homogenizing project of development. Difference, instead of leading to the richness of diversity, becomes the basis for division and an ideology of separation.

Globalization III: 'Free Trade'

Globalization and homogenization are now being carried out not by nation states, but by global powers that control global markets. 'Free trade' is the ruling metaphor for globalization in our times. But far from protecting the freedom of citizens and countries, free trade negotiations and treaties have become the primary locations for the use of coercion and force. The Cold War era has ended, the era of trade wars has begun.

Among the exemplars of violence in the free trade era is the U.S. Trade Act, especially the Super and Special 301 clauses that allow the United States to take unilateral action against any country that does not open up its market to U.S. corporations. Super 301 forces freedom for investment; Special 301 forces freedom for monopoly control of markets through intellectual property rights protection. Free trade is, in fact, an asymmetric arrangement that combines liberalization and protectionism for Western interests. As Martin Khor has said: "Free trade and liberalization were only nice slogans waved to move the [Uruguay] Round forward. The reality was 'liberalization if it benefits us, protectionism if it benefits us, what counts is our self-interest'." [9]

Third World countries had resisted the expansion of GATT into new areas like services, investments, and intellectual property rights. By merely affixing the phrase 'trade-related' to issues that are decided domestically, GATT, through the World Trade Organization, will not merely regulate international trade, but in essence will determine domestic policy.

This brute force continued to be used against the Third World even in the multilateral negotiations of the Uruguay Round of GATT. In a speech, Fernando Jaramillo, Chair of the Group of 77 and Colombia's permanent representative to the United Nations, said, "The Uruguay Round is proof again the developing world continues to be sidelined and rejected when it comes to defining areas of vital importance to their survival." [10]

The very process itself is undemocratic and unilateral. Free trade treaties like GATT are forced on citizens and weaker trading partners, such as Third World countries. In 1991, for instance, a take-it-or-leave-it draft was prepared by GATT Secretary General Arthur Dunkel; a draft which in India has taken on the not so pleasant acronym of DDT (Dunkel Draft Text). An even more blatant example is the last stage of GATT negotiations in December 1993, in which two men—Micky Kantor, the U.S. trade representative, and Leon Brittan, the negotiator for the European Community sat behind closed doors and then presented the world with a 'free trade' treaty.

Despite insisting that the negotiations were global, the countries of the North refused in the end to accept any discussions, even bilaterally, with the countries of the Third World. This is neither multilateralism nor global democracy.

A new authoritarian structure emerges, as Ambassador Jaramillo observes:

> The Bretton Woods Institutions continue to be made the centre of gravity for the principal economic decisions that affect the developing countries. We have all been witnesses to the conditionalities of the World Bank and the IMF. We all know the nature of the decisionmaking system in such institutions; their undemocratic character, their lack of transparency, their dogmatic principles, their lack of pluralism in the debate of ideas, and their impotence to influence the policies of the industrialized countries.
>
> This also seems to be applicable to the new World Trade Organization. The terms of its creation suggest that this organization will be dominated by the industrialized countries and that its fate will be to align itself with the World Bank and IMF.
>
> We could announce in advance the birth of a New Institutional Trinity which would have as its specific function to control and dominate the economic relations that commit the developing world.[11]

In reality, free trade has vastly expanded the freedom and powers of transnational corporations to trade and invest in most countries of the world, while significantly reducing the powers of national governments in order to restrict their operations. Multinational corporations, the real power in the Uruguay Round, have gained new rights and given up old obligations to protect workers' rights and the environment.

Free trade is not free; it protects the economic interests of the powerful transnational corporations, which already control 70 percent of the world's trade and for whom international trade is an imperative. Transnational corporate freedom is based on the destruction of citizens' freedom everywhere, and the little remnants of independence that the Third World had after the last two waves of colonization. In essence, GATT cripples the democratic institutions of individual countries—local councils, regional governments, and parliaments—leaving them unable to carry out the will of their citizens.

While GATT might increase the volume of internationally traded goods and services, it will also increase unemployment and generate scarcity for those excluded from the global economy. The Indian Commerce Minister admitted in 1994 that unemployment in India will increase dramatically as a result of GATT. In Germany, the unemployment rate is expected to go up from 7.4 to 11.3 percent. France is moving from 9.5 to 12.1 percent, Britain from 9.7 to 10.4 percent. The top 1,000 British companies shed 1.5 million jobs in one year. Their total workforce dropped from 8.6 million to just over 7 million. The French Assembly anticipates that French unemployment will rise by 3.5 million in the next 10 years. According to Jeremy Rifkin in his book *The End of Work*, in the United States, 90 million jobs out of a total of 120 million are vulnerable to displacement by the restructuring of production.[12] A recent *Wall Street Journal* article projects that 1.5 to 2.5 million American jobs could be lost each year for the foreseeable future.

Countries are also reducing security benefits for workers. France announced a pension freeze; Germany reduced unemployment benefits. A leaked U.K. government document suggests plans to deregulate worker health and safety regulations. These range from ending the requirement for employers to provide toilet paper and soap at work to the partial ending of controls on industrial hazards.

Instead of protecting workers' rights domestically, and instead of ending the structural adjustment policies of the World Bank that lower Third World wages, the industrialized countries now argue that low wages in the Third World lead to 'social dumping' in international trade and that trade sanctions are necessary to protect rich countries.

Hundreds of millions of farmers' livelihoods around the world are under threat from GATT and the new biotechnologies. The 'producer retirement' programmes in the agriculture treaty are basically a displacement policy for farmers. In addition, monopoly control of seeds and plant varieties further add to the displacement pressures on the small farmers of the Third World who are the original breeders and custodians of plant genetic resources.

In response to the violence of free trade, its victims will react. For example, the January 1, 1994 revolt of the Zapatistas in the Chiapas region of Mexico, in the year that coincided with the beginning of the North American Free Trade Agreement, cost 107 lives. According to a rebel leader: "The free trade agreement is a death certificate for the Indian peoples of Mexico." Inspired by the Chiapas rebellion, other groups in Mexico are coming out in protest. As the leader of the National Coalition of Indigenous Peoples said: "Don't test us, because Zapatistas could appear all over the country."

The structural adjustment programmes of the IMF and World Bank, which tried to establish free trade in the pre-GATT era, indicate the three levels of violence created by the third wave of globalization.

First, there is the violence of the structural adjustment programmes themselves, which rob people of food, health care, and education.

When people's very survival is threatened, they protest to protect their rights. These protests, in turn, face repression from regimes committed to the structural adjustment conditionalities of the World Bank and IMF. A Peruvian economist has estimated that in the several outbreaks of protest against structural adjustment programmes, nearly 3,000 people have died, 7,000 have been wounded, and 15,000 have been arrested.

Finally, the economic and political vulnerability created by robbing people of their self-organizing, self-governing, and self-provisioning capacities also creates conditions for engineered violence, in which vested interests organize vulnerable groups along ethnic and religious lines to declare war on each other. No continent is free of such civil wars, engineered along the fractures of racial, religious, or ethnic differences. The end of the Cold War has, in fact, seen war introduced on a global scale in civil society. Diversity has been transformed into a problem in a globalizing and homogenizing world.

The experiences of Somalia and Rwanda are vivid illustrations of the manifold violence of globalization.

The Somalian crisis has been interpreted as a residue of 'tribalism'. According to Michel Chossudovsky, however, the civil war in Somalia is more intimately connected to the effects of globalization in the form of structural adjustment programmes. Somalia had a pastoral economy based on exchange between nomadic herdsmen and small peasants; it remained virtually self-sufficient in terms of food. Livestock made up 80 percent of Somalia's export earnings until 1983.

The IMF-World Bank adjustment programmes in the 1980s destroyed Somalia's economic and social fabric. Devaluation and liberalization of imports led to an erosion of domestic agricultural production. Food aid increased 15-fold between the mid-'70s and mid-'80s, leading to the displacement of farmers. Privatization of veterinary services and water resources led to a collapse of the livestock sector. As Chossudovsky reports:

> The IMF-World Bank program has led the Somali economy into a vicious circle: the decimation of the herds pushed the nomadic pastoralists into starvation which in turn backlashed on grain producers who sold or bartered their grain for cattle. The entire social fabric of the pastoralist economy was undone. The collapse in foreign exchange earnings from declining cattle exports and remittances backlashed on the balance of payments and the state's public finances leading to the breakdown of the government's economic and social programme.[13]

The Rwandan genocide had similar links to the globalization processes of structural adjustment. In 1989, the International

Coffee Agreement reached a deadlock, and worldwide coffee prices plunged by more than 50 percent. Rwanda's export earnings from coffee declined by 50 percent between 1987 and 1991.

In November 1990, a 50 percent devaluation of the Rwandan franc was carried out under the World Bank–IMF adjustment programme. The balance of payments situation deteriorated dramatically, and the outstanding external debt, which had already doubled since 1985, increased by another 34 percent between 1989 and 1992. In June 1992, another devaluation was ordered, leading to a 25 percent decrease in coffee production. Chossudovsky explains:

> The crisis of the coffee economy backlashed on the production of cassava, beans and sorghum. The system of savings and loan cooperatives which provided credit to small farmers had also disintegrated. Moreover, with the liberalization of trade and the deregulation of grain markets as recommended by the Bretton Woods Institutions, heavily subsidized cheap food imports and food aid from the rich countries were entering Rwanda with the effect of destabilizing local markets.[14]

Everywhere, globalization leads to the destruction of local economies and social organization, pushing people into insecurity, fear, and civil strife. The violence against people's livelihoods builds up into the violence of war.

There is only one way to contain these epidemics of violence. We must, with sensitivity and responsibility, wherever and whoever we are, once again make peace with diversity. We have to learn that diversity is not a recipe for conflict or chaos, but is our only chance for a more sustainable and just future in social, political, economic, and environmental terms. It is our only means to survival.

Chapter Seven

Nonviolence and Cultivation of Diversity

An intolerance of diversity is the biggest threat to peace in our times; conversely, the cultivation of diversity is the most significant contribution to peace—peace with nature and between diverse peoples. The cultivation of diversity has to be a conscious and creative act, intellectually and in practice. It demands more than mere tolerance of diversity, because tolerance alone is not enough to contain the wars unleashed by the intolerance of difference.

Diversity is intimately linked to the possibility of self-organization. Decentralization and local democratic control are political corollaries of the cultivation of diversity. Peace is also derived from conditions in which diverse species and communities have the freedom to self-organize and evolve according to their own needs, structures and priorities.

Globalization has undermined the conditions for self-rule, self-governance, and self-organization. It has established a violent order, both in terms of the coercive structures needed to maintain the order, and of the ecological and social disintegration and violence that are products of that order.

The cultivation of diversity involves reclaiming the right to self-organize for those coerced into living by imposed measures. For the dominant groups of nations and humans, who impose their priorities and patterns on the living diversity of peoples and other species, the cultivation of diversity involves seeing the capacity and intrinsic value of the 'other'—other cultures and other species. It involves giving up the will to control, an imperative rooted in the fear of that which is free, a fear that gives rise to violence. The cultivation of diversity is, therefore, a non-violent response to the violence of globalization, homogenization, and monocultures.

Biodiversity is fast becoming the primary site of conflict between worldviews based on diversity and non-violence and those based on monocultures and violence.

Biodiversity has been seen as the exclusive domain of conservationists. Yet nature's diversity converges with cultural diversity. Different cultures have emerged in accordance with different endowments of species in varied ecosystems. They have found diverse ways to conserve and utilize the rich biological wealth of their habitats. New species have been introduced into their ecosystems with careful experimentation and innovation. Biodiversity does not merely symbolize nature's richness; it embodies diverse cultural and intellectual traditions.

There are two conflicting paradigms of biodiversity. The fist paradigm is held by local communities, whose survival and sustenance is linked to the utilization and conservation of biodiversity. The second is held by commercial interests, whose profits are linked to utilizing global biodiversity as inputs for large-scale, homogeneous, centralized, and global production systems. For local indigenous communities, conserving biodiversity means conserving their rights to their resources, knowledge, and production systems. For commercial interests, such as pharmaceutical and agricultural biotechnology companies, biodiversity in itself has no value; it is merely raw material. Production is based on biodiversity destruction, as local production systems based on diversity are displaced by production based on uniformity.

The conflict between these two paradigms is exacerbated by the emergence of new biotechnologies for the manipulation of life and new legal regimes for the monopoly control on life.

Both the technological and legal trends are toward monocultures and uniformity. They are predicated on wiping out diverse technological options as well as the pluralistic ways people have related to nature and evolved systems of rights and obligations. The monopolizing control of the molecular monoculture mind is most powerful through the rise of the new tools of genetic engineering. As Jack Kloppenburg has warned:

> Though the capacity to move genetic material between species is a means for introducing additional variation, it is also a means for engineering uniformity across species.[1]

The production of transgenic species has been achieved through the crossing of species boundaries, which have been nature's way of maintaining distinctiveness and diversity. While the ecological impact of crossing these boundaries has not yet been fully anticipated or assessed, a few predictions are possible. For example, breeding plants for herbicide resistance is one of the largest areas of investment in agricultural biotechnology. The aim is to concentrate market control of agriculture into the hands of a few corporations. At the same time, however, it introduces new pressures for uniformity since crops not resistant to these herbicides cannot be grown in fields contaminated by their excessive use. Further, in regions of biodiversity, the introduction of crops genetically engineered for herbicide tolerance can end up creating super-weeds, as genes for herbicide resistance transfer to weedy relatives of crops.

From an ecological perspective, these technological options are wasteful, hazardous, and unnecessary. They are being spread not only because legal systems create conditions of monopoly control over biological material and markets through intellectual property rights. Like patents, IPRs are supposed to be rights to products of the mind. Yet different cultures have evolved different knowledge traditions, and different values and norms for the sharing and exchange of that knowledge. Thus, for example, at the beginning of the agricultural season in India, during a festival called Akti, farmers bring their diverse seeds together and exchange them. In this cultural context, the seed is treated as common, not private, property. Intellectual property rights, however, are based on a knowledge monoculture that excludes diverse knowledge traditions. IPRs colonize the intellectual heritage of non-Western cultures as well as their natural heritage, which is concentrated in what have become Third World countries over five centuries of unilaterally determined exchange.

The TRIPs treaty in GATT recognizes IPRs only as private, not common, rights. This excludes all kinds of knowledge, ideas, and innovations that take place in the intellectual commons—in villages among farmers, in forests among tribespeople, and even in universities among scientists. Such IPR protection will stifle the pluralistic ways of knowing that have enriched our world.

IPRs are recognized only when knowledge and innovation generate profits, not when they meet social needs. Profits and capital accumulation are the only ends to which creativity is put; the social good is no longer recognized.

The universalization of the preferred priorities of a very small part of human society will destroy creativity, not encourage it. By reducing human knowledge to the status of private property, intellectual property rights shrink the human potential to innovate and create; they transform the free exchange of ideas into theft and piracy.

In reality, IPRs are the sophisticated name for modern piracy. With no regard or respect for other species and cultures, IPRs are a moral, ecological, and cultural outrage. Moreover, IPR actions in the biodiversity domain are tainted with cultural, racial, and species prejudice and arrogance.

The GATT is the platform where the capitalistic, patriarchal notion of freedom as the unrestrained right of men with economic power to own, control, and destroy life is articulated as free trade. But for the Third World, and particularly for women, freedom has different meanings. In what seems the remote domain of international trade, these different meanings of freedom are a focus of contest and conflict. Free trade in food and agriculture is the concrete location of the most fundamental ethical and economic issues facing humans today.

The biodiversity issue is an opportunity to recover diversity at the ethical, ecological, epistemological, and economic levels.

The conservation of biodiversity, at the most fundamental level, is the ethical recognition that other species and cultures have rights, that they do not merely derive value from economic exploitation by a few privileged humans. The patenting and ownership of life forms is ethically a statement of the opposite belief.

Biodiversity conservation is a product of the cultural contributions of communities that respect other species, and that have evolved the knowledge of diverse species and their interactions to allow for a utilization in harmony with the objectives of conservation.

Conservation of biodiversity, therefore, involves the conservation of cultural diversity and a plurality of knowledge traditions. This plurality, in turn, is ecologically necessary for survival in times of rapid change and accelerated breakdown.

Even as the world becomes more and more uncertain and unpredictable, technological and economic models are being based on a linear paradigm that assumes total certainty and control. While we live with the negative social and ecological consequences of past systems of centralization and uniformity in production, the centralization and uniformity is being increased.

It is often assumed that centralization and uniformity are growth imperatives. But for what kind of growth?

When multidimensional, diverse systems are perceived in their entirety, they are found to have high productivity. Their 'low productivity' is the product of an approach that evaluates and assesses within a one-dimensional framework, which is, in turn, related to an instrumental worldview. When a pig or cow is simply treated as a bioreactor, for instance, to produce a certain kind of chemical for the pharmaceutical industry, it can be re-engineered and redesigned without any ethical constraint. Diversity as a worldview allows diverse components to be perceived, irrespective of their size. The recognition of the diverse roles and interdependence of each part puts limits on our exploitation of other species, and limits human arrogance.

Navdanya (nine seeds) or *barnaja* (twelve crops) are examples of highly productive systems of mixed farming or polycultures based on diversity, yielding more than any monoculture can. Unfortunately, they are disappearing—not because of their low productivity, but because they need no inputs, being based on symbiosis with legumes providing nitrogen to cereals. In addition, their outputs are diverse—providing all of the nutritional inputs a family needs. This diversity, however, acts against commercial interests, which need to maximize the production of a single output to maximize profits. Polycultures, by their very nature, are ecologically prudent. Thus recovering diversity in production provides a countervailing force to the globalized, centralized, and homogeneous systems of production that are destroying livelihoods, cultures, and ecosystems everywhere.

By pluralizing our options, we simultaneously create the tools for reconstruction and resistance. In India, a massive movement—the Seed *Satyagraha*—has emerged over the past few years in response to the threats of recolonization through GATT, especially its intellectual property rights clauses. According to Gandhi, no tyranny can enslave a people who consider it immoral to obey laws that are unjust. As he stated in *Hind Swaraj*:

> As long as the superstition that people should obey unjust laws exists, so long will slavery exist. And a passive resister alone can remove such a superstition.[2]

Satyagraha is the key to self-rule, or *swaraj*. The phrase that echoed most during India's freedom movement was "*Swaraj hamara janmasidh adhikar hai*" (self-rule is our birthright). For Gandhi, and for the contemporary social movements in India, self-rule did not imply governance by a centralized state, but by decentralized communities. "*Nate na raj*" (our rule in our village) is one slogan from India's grassroots environmental movement.

At a massive rally in Delhi in March 1993, a charter of farmers' rights was developed. One of the rights is local sovereignty. Local resources have to be managed on the principle of local sovereignty, wherein the natural resources of the village belong to that village.

A farmer's right to produce, exchange, modify, and sell seed is also an expression of *swaraj*. Farmers' movements in India have declared they will violate the GATT treaty, if it is implemented, since it violates their birthright

Another Gandhian concept that the Seed *Satyagraha* has revived is that of *swadeshi*. *Swadeshi* is the spirit of regeneration, a method of creative reconstruction. According to the *swadeshi* philosophy, people already possess, both materially and morally, what they need to free themselves of oppressive structures.

Swadeshi, for Gandhi, was a positive concept based on building on the resources, skills, and institutions of a community, and when necessary, transforming them. Imposed resources, institutions, and structures leave a people unfree. For Gandhi, *swadeshi* was central to the creation of peace and freedom.

In the free trade era, the rural communities of India are redefining non-violence and freedom by reinventing the concepts of *swadeshi*, *swaraj*, and *satyagraha*. They are saying 'no' to unjust laws, like the GATT treaty, that legalize the theft of the biological and intellectual heritage of Third World communities.

A central part of the Seed *Satyagraha* is to declare the common intellectual rights of Third World communities. While the innovations of Third World communities might differ in process and objectives from those in the commercial world of the West, they cannot be discounted just because they are different. The knowledge of the rich bounties of nature's diversity has been a gift from the Third World. But Seed *Satyagraha* has gone beyond just saying 'no'. It has created alternatives by building community seed banks, strengthening farmers' seed supply, and searching for sustainable agricultural options suitable for different regions.

The seed has become the site and symbol of freedom in the age of manipulation and monopoly of its diversity. It plays the role of Gandhi's spinning wheel in this period of recolonization through free trade. The *charkha* (spinning wheel) became an important symbol of freedom not because it was big and powerful, but because it was small; it could come alive as a sign of resistance and creativity in the smallest of huts and poorest of families. In smallness lay its power.

The seed, too, is small. It embodies diversity and the freedom to stay alive. And seed is still the common property of small farmers in India. In the seed, cultural diversity converges with biological diversity. Ecological issues combine with social justice, peace, and democracy.

Notes

Introduction

1. John Locke, *Two Treatises of Government*, ed. Peter Caslett (Cambridge University Press, 1967).
2. John Winthrop 'Life and Letters', quoted in Djelal Kadir, *Columbus and the Ends of the Earth* (Berkeley: University of California Press, 1992), p. 171.

Chapter One

1. Vandana Shiva, *Monocultures of the Mind* (London: Zed Books, 1993).
2. Robert Sherwood, *Intellectual Property and Economic Development* (Boulder, San Francisco, and Oxford: Westview Press).
3. Ibid., pp. 196–97.
4. Emanuel Epstein, quoted in Kenneth Martin, *Biotechnology: The University-Industrial Complex* (New Haven and London: Yale University Press), pp. 109-10.
5. Martin Kenny, quoted in *Biotechnology: The University-Industrial Complex.*
6. Charles Darwin, *The Formation of Vegetable Mould Through the Action of Worms with Observations of Their Habits* (London: Murray, 1891).
7. David Ehrenfeld, *Beginning Again* (New York and Oxford: Oxford University Press, 1993), pp. 70–71.

Chapter Two

1. Andrew Kimbrell, *The Human Body Shop* (New York: HarperCollins Publishers, 1993).
2. Ibid.
3. Key Dismukes, quoted in Brian Belcher and Geoffrey Hawtin, *A Patent on Life: Ownership of Plant and Animal Research* (Canada: IDRC, 1991).
4. Vandana Shiva, *Monocultures of the Mind* (London: Zed Books, 1993).
5. Rural Development Foundation International Communique, Ontario, Canada, June 1993.
6. *New Scientist*, January 9, 1993.
7. Carolyn Merchant, *The Death of Nature: Women, Ecology and the Scientific Revolution* (New York: Harper & Row, 1980), p. 182.
8. Robert Wesson, *Beyond Natural Selection* (Cambridge, MA: The MIT Press, 1993), p. 19.
9. J. W. Pollard, 'Is Weismann's Barrier Absolute?', in eds. M. W. Ho and P. T. Saunders, *Beyond Neo-Darwinism: Introduction to the New Evolutionary Paradigm* (London: Academic Press, 1984), pp. 291-315.
10. Francis Crick, 'Lessons from Biology', *Natural History 97* (November 1988): 109.
11. Richard Lewontin, *The Doctrine of DNA* (Penguin Books, 1993).
12. Beyond Natural Selection, p. 29.
13. Lily E. Kay, *The Molecular Vision of Life: Caltech, The Rockefeller Foundation and the Rise of the New Biology* (Oxford, England: Oxford University Press, 1993), p. 6.
14. Ibid., p. 8.
15. *The Doctrine of DNA*, p. 22.
16. *The Molecular Vision of Life*, pp. 8-9 (see ref. 13 above).

17. Roger Lewin, 'How Mammalian RNA Returns to Its Genome', *Science* 219 (1983): 1052-1054.
18. Richard Dawkins, *The Selfish Gene* (Oxford, England: Oxford University Press, 1976).
19. Humberto R. Maturana and Francisco J. Varela, *The Tree of Knowledge: The Biological Roots of Human Understanding* (Boston, MA: Shambhala Publications, 1992).
20. L. J. Taylor, quoted in David Coats, *Old McDonald's Factory Farm* (NY: Continuum, 1989), p. 32.
21. *Beyond Natural Selection* (see ref. 8 above).
22. Mae Wan Ho, 'Food, Facts, Fallacies and Fears' (paper presented at National Council of Women Symposium, United Kingdom, March 22, 1996).
23. Vandana Shiva, *et al.*, *Biosafety* (Penang: Third World Network, 1996).
24. Phil J. Regal, 'Scientific Principles for Ecologically Based Risk Assessment of Transgene Organisms', *Molecular Biology*, Vol. 3 (1994): 5-13.
25. Elaine Ingham and Michael Holmes, 'A note on recent findings on genetic engineering and soil organisms', 1995.
26. R. Jorgensen and B. Anderson, 'Spontaneous Hybridization Between Oilseed Rape (*Brassica Napas*) and Weedy *B. campestris* (Brassicaceae): A Risk of Growing Genetically Modified Oilseed Rape', *American Journal of Botany* (1994).
27. Rural Development Foundation International Communique, United States, July/August 1996, pp. 7-8.
28. 'Pests Overwhelm Bt. Corron Crop', *Science* 273: 423.
29. Rural Development Foundation International Communique, United States, July/August 1996, pp. 7-8.
30. The Battle of the Bean, 'Splice of Life' (October 1966).

Chapter Three

1. Johann Jacob Bachofen, quoted in Marta Weigle, *Creation and Procreation* (Philadephia: University of Pennsylvania Press, 1989).
2. Ibid.
3. Claudia Von Werlhof, 'Women and Nature in Capitalism', in ed. Maria Mies, *Women: The Last Colony* (London: Zed Books, 1989)
4. John Pilger, *A Secret Country* (London: Vintage, 1989).
5. Ibid.
6. Ibid.
7. Carolyn Merchant, *The Death of Nature: Women, Ecology and the Scientific Revolution* (New York: Harper & Row, 1980).
8. Vandana Shiva, *The Violence of the Green Revolution* (Penang: Third World Network, 1991).
9. Jack Kloppenburg, *First the Seed* (Cambridge, England: Cambridge University Press, 1988).
10. Quoted in Jack Doyle, *Altered Harvest* (New York: Viking, 1985), p. 310.
11. *First the Seed*, p. 185 (see ref. 9 above).
12. Stephen Witt, 'Biotechnology and Genetic Diversity', California Agricultural Lands Project, San Francisco, CA, 1985.
13. Pat Mooney, 'From Cabbages to Kings', in *Development Dialogue* (1988): 1-2 and 'Proceedings of the Conference on Patenting of Life Forms' (Brussels: ICDA, 1989).
14. 'Biotechnology and Genetic Diversity'.

15. *Altered Harvest* (see ref. 10 above).
16. Hans Leenders, 'Reflections on 25 Years of Service to the International Seed Trade Federation', *Seedsmen's Digest* 37: 5, p. 89.
17. Quoted in *First the Seed*, p. 266.
18. *First the Seed*, p. 266.
19. Rural Advancement Foundation International, *Biodiversity, UNICED and GATT*, Ottawa, Canada, 1991.
20. Ibid.
21. Ibid.
22. Food and Agriculture Organization (FAO), International Undertaking on Plant Genetic Resources, DOC C83/II REP/4 and 5, Rome, Italy, 1983.
23. Keystone International Dialogue on Plant Genetic Resources, Final Consensus Report of Third Plenary Session, Keystone Center, Colorado, May 31-June 4, 1991.
24. Genetic Resources Action International (GRAIN), 'Disclosures: UPOV sells out', Barcelona, Spain, December 2, 1990.
25. Vandana Shiva, 'Biodiversity, Biotechnology and Bush', *Third World Network Earth Summit Briefings* (Penang: Third World Network, 1992).
26. Vandana Shiva, 'GATT and Agriculture', *The* [Bombay] *Observer* (1992).
27. Neil Postman, *Technology: The Surrender of Culture to Technology* (A. Knopf, 1992).
28. Peter Singer and Deane Wells, *The Reproductive Revolution: New Ways of Making Babies* (Oxford, England: Oxford University Press, 1984).
29. Phyllis Chesler, *Sacred Bond: Motherhood Under Siege* (London: Virago, 1988).
30. European Patent Office, application no. 833075534.
31. 'Women and Nature in Capitalism' (see ref. 3 above).
32. Marilyn Waring, *If Women Counted* (New York: Harper & Row, 1988).
33. United Nations Conference on Environment and Development, 'Agenda 21', adopted by the plenary on June 14, 1992, published by the UNCED Secretariat, Conches, Switzerland.

Chapter Four

1. Vandana Shiva, *Monocultures of the Mind* (London: Zed Books, 1993).
2. *Charaka Samhita*, Sutra Sthaana, Chap. 1, Sloka, pp. 120-21.
3. R. Stone, 'A Biocidal Tree Begins to Blossom', *Science* (February 28, 1992).
4. Letter to Professor Narjundaswamy, convener of the Karnataka Rajya Raitha Sangha Farmers' Organization.
5. The EPA does not accept the validity of traditional knowledge and has imposed a full series of safety tests upon one of the products, Margosan-O.
6. World Resources Institute, 1993.
7. Susan Laird, 'Contracts for Biodiversity Prospecting', in *Biodiversity Prospecting*, World Resource Institute (1994): 99.
8. Farnsworth, quoted in *Biodiversity Prospecting* (1990): 119.
9. SCRIP, quoted in *Biodiversity Prospecting* (1992): 102-3.
10. *Biodiversity Prospecting* (1991): 103.
11. James Enyart, 'A GATT Intellectual Property Code', *Less Nouvelles* (Tune 1990): 54-56.
12. 'Basic Framework for GATT Provisions on Intellectual Property', statement of views of the European, Japanese, and U.S. business communities, June 1988.
13. Ibid.
14. Ibid.
15. Ibid.

Chapter Five

1. Jack Doyle, *Altered Harvest* (New York: Viking, 1985), p. 256.
2. Ibid.
3. Peter Wheale and Ruth McNally, 'Genetic Engineering: Catastrophe of Utopia', *U.K. Harvester* (1988): 172.
4. U.S. Academy of Sciences, *Field Testing Genetically Modified Organisms: Framework for Decisions* (Washington, D.C.: National Academy Press, 1989).
5. Anne Kapuscinski and E. M. Hallerman, *Canadian Journal of Fisheries and Aquatic Sciences*, Vol. 48 (1991): 99–107.
6. G. B. Pant, 'The Forest Problem in Kumaon', *Gyanodaya Prakashan* (1922): p. 75.
7. G. S. Halappa, *History of Freedom Movement in Karnataka*, Vol. II (Bangalore: Government of Mysore, 1969), p. 175.
8. 'National Conservation Strategy Action Plan for the National Policy on Natural Resources and the Environment', National Conservation Strategy Secretariat, Addis Ababa, Vol. II (December 1994): 7.

Chapter Six

1. Vandana Shiva, *The Violence of the Green Revolution* (London: Zed Books, 1991), p. 89.
2. Vandana Shiva, 'Resources', in ed. Wolfgang Sachs, *Development Dictionary* (London: Zed Books, 1992), p. 206.
3. Rabindranath Tagore, 'Tapovan' (Hind)), Tikamagarh, Gandhi Bhavan, undated, pp. 1-2.
4. Robert Boyle, quoted in Brian Easlea, *Science and Sexual Oppression: Patriarchy's Confrontation with Woman and Nature* (London: Weidenfeld and Nicholson, 1981), p. 64.
5. A. W. Crosby, *The Colombian Exchange* (Westport, CT: Greenwood Press, 1972), p. 12.
6. Basil Davidson, *Africa in History* (New York Collier Books, 1974), p. 178.
7. *The Violence of the Green Revolution*, p. 171 (see ref. 1 above).
8. Francine Frankel, *The Political Challenge of the Green Revolution* (Princeton, NJ: Princeton University, 1972), p. 38.
9. Martin Khor, *The Uruguay Round and Third World Sovereignty* (Penang: Third World Network, 1990), p. 29.
10. Quoted in Chakravarthi Raghavan, 'A Global Strategy for the New World Order', *Third World Economics*, No. 81/82 (January 1995).
11. Ibid.
12. Jeremy Rifkin, *The End of Work* (New York: Tarcher/Putnam, 1994).
13. Michel Chossudovsky, 'Global Poverty', unpublished manuscript.
14. Ibid.

Chapter Seven

1. Jack Kloppenburg, *First the Seed* (Cambridge University Press, 1988).
2. M. K. Gandhi, *Hind Swaraj or Indian Home Rule* (Ahmedabad: Navjivan Publishing House, 1938), p. 29.

Index

The Gaia Foundation

The Gaia Foundation's work is guided by the priorities and requests of its Associates, a key one of whom is Vandana Shiva. The Associates are policy-makers, NGOs, specialists, scientists and grassroot fieldworkers and innovators from Asia, Africa and South America, dedicated to the protection of cultural and biological diversity and the genetic resources fundamental to food and medicine. They work at many levels and help to bring the concerns and perspectives of grassroots communities to government policy-makers and international arenas such as the Convention on Biological Diversity and the World Trade Organisation.

Several Associates have requested Gaia to focus on two major threats to biodiversity: patents and genetic engineering. It is therefore a great pleasure to be able to co-publish Vandana Shiva's clear and eloquent book *Biopiracy* with Green Books.

The Gaia Foundation focuses on building and extending networks and coalitions. This is central to the work of assuming and sharing responsibility and democratizing the way decisions are made on these critical issues. It is also vital to empower people to create the institutions and governments which we will need as we enter the next millennium. We must protect the planet's biological and cultural wealth, and resist globalization and the imposition of 'free' trade. To do this effectively, we need collective rights systems to defend the accumulated wisdom and innovation of those who have enhanced and protected it through the generations until today.

The Gaia Foundation, 18 Well Walk, London NW3 1LD
Phone: 44 171 435 5000 Fax: 44 171 431 0551
Email: gaia@gaianet.org

Also from Green Books:

Global Spin: The Corporate Assault on Environmentalism
by Sharon Beder

"The most complete study so far of the million dollar propaganda machine the transnational corporations have built up to discredit the environmental movement" - Edward Goldsmith

For a copy of our latest catalogue, contact us at:

Green Books, Foxhole, Dartington, Totnes, Devon TQ9 6EB
Phone: 01803 863260 Fax: 01803 863843
greenbooks@gn.apc.org www.greenbooks.co.uk